Advances in Nanoparticles

Advances in Nanoparticles

Synthesis, Characterization, Theoretical Modelling, and Applications

Special Issue Editor

Luca Pasquini

MDPI • Basel • Beijing • Wuhan • Barcelona • Belgrade • Manchester • Tokyo • Cluj • Tianjin

Special Issue Editor
Luca Pasquini
University of Bologna
Italy

Editorial Office
MDPI
St. Alban-Anlage 66
4052 Basel, Switzerland

This is a reprint of articles from the Special Issue published online in the open access journal *Nanomaterials* (ISSN 2079-4991) (available at: https://www.mdpi.com/journal/nanomaterials/special_issues/advance_nanoparticle).

For citation purposes, cite each article independently as indicated on the article page online and as indicated below:

LastName, A.A.; LastName, B.B.; LastName, C.C. Article Title. *Journal Name* **Year**, *Article Number*, Page Range.

ISBN 978-3-03928-582-2 (Pbk)
ISBN 978-3-03928-583-9 (PDF)

Cover image courtesy of Luca Pasquini.

© 2020 by the authors. Articles in this book are Open Access and distributed under the Creative Commons Attribution (CC BY) license, which allows users to download, copy and build upon published articles, as long as the author and publisher are properly credited, which ensures maximum dissemination and a wider impact of our publications.
The book as a whole is distributed by MDPI under the terms and conditions of the Creative Commons license CC BY-NC-ND.

Contents

About the Special Issue Editor vii

Preface to "Advances in Nanoparticles" ix

Zhixia Zhang, Chunjin Wei, Wenting Ma, Jun Li, Xincai Xiao and Dan Zhao
One-Step Hydrothermal Synthesis of Yellow and Green Emitting Silicon Quantum Dots with Synergistic Effect
Reprinted from: *Nanomaterials* **2019**, *9*, 466, doi:10.3390/nano9030466 1

Nicola Patelli, Andrea Migliori, Vittorio Morandi and Luca Pasquini
One-Step Synthesis of Metal/Oxide Nanocomposites by Gas Phase Condensation
Reprinted from: *Nanomaterials* **2019**, *9*, 219, doi:10.3390/nano9020219 15

Inderjeet Singh and Balaji Birajdar
Effective La-Na Co-Doped TiO_2 Nano-Particles for Dye Adsorption: Synthesis, Characterization and Study on Adsorption Kinetics
Reprinted from: *Nanomaterials* **2019**, *9*, 400, doi:10.3390/nano9030400 31

Hokuto Fuse, Naoto Koshizaki, Yoshie Ishikawa and Zaneta Swiatkowska-Warkocka
Determining the Composite Structure of Au-Fe-Based Submicrometre Spherical Particles Fabricated by Pulsed-Laser Melting in Liquid
Reprinted from: *Nanomaterials* **2019**, *9*, 198, doi:10.3390/nano9020198 47

Jian Yu, Tingting Xiao, Xuemin Wang, Xiuwen Zhou, Xinming Wang, Liping Peng, Yan Zhao, Jin Wang, Jie Chen, Hongbu Yin and Weidong Wu
A Controllability Investigation of Magnetic Properties for FePt Alloy Nanocomposite Thin Films
Reprinted from: *Nanomaterials* **2019**, *9*, 53, doi:10.3390/nano9010053 59

Liyao Zhang, Yuxin Song, Qimiao Chen, Zhongyunshen Zhu and Shumin Wang
InPBi Quantum Dots for Super-Luminescence Diodes
Reprinted from: *Nanomaterials* **2018**, *8*, 705, doi:10.3390/nano8090705 69

Genli Shen, Mi Liu, Zhen Wang and Qi Wang
Hierarchical Structure and Catalytic Activity of Flower-Like CeO_2 Spheres Prepared Via a Hydrothermal Method
Reprinted from: *Nanomaterials* **2018**, *8*, 773, doi:10.3390/nano8100773 79

Jin Wang, Xuemin Wang, Jian Yu, Tingting Xiao, Liping Peng, Long Fan, Chuanbin Wang, Qiang Shen and Weidong Wu
Tailoring the Grain Size of Bi-Layer Graphene by Pulsed Laser Deposition
Reprinted from: *Nanomaterials* **2018**, *8*, 885, doi:10.3390/nano8110885 87

Chenying Wang, Xidong Ren, Yujie Su and Yanjun Yang
Application of Glycation in Regulating the Heat-Induced Nanoparticles of Egg White Protein
Reprinted from: *Nanomaterials* **2018**, *8*, 943, doi:10.3390/nano8110943 95

Yanhua Yao, Nannan Zhang, Xiao Liu, Qiaofeng Dai, Haiying Liu, Zhongchao Wei, Shaolong Tie, Yinyin Li, Haihua Fan and Sheng Lan
A Novel Fast Photothermal Therapy Using Hot Spots of Gold Nanorods for Malignant Melanoma Cells
Reprinted from: *Nanomaterials* **2018**, *8*, 880, doi:10.3390/nano8110880 107

Bianca Moldovan, Vladislav Sincari, Maria Perde-Schrepler and Luminita David
Biosynthesis of Silver Nanoparticles Using *Ligustrum Ovalifolium* Fruits and Their Cytotoxic Effects
Reprinted from: *Nanomaterials* **2018**, *8*, 627, doi:10.3390/nano8080627 . . . . **121**

About the Special Issue Editor

Luca Pasquini received a Ph.D. in Condensed Matter Physics at the University of Bologna. He worked as a post-doc at the University of Stuttgart, the ESRF synchrotron, and the University of Bologna, where he is currently Associate Professor of experimental materials physics. He has over 20 years of experience on the growth of nanoparticles and nanostructures using physical methods and on the investigation of structure–property relationships. He is author of about 110 papers in international journals. His current research interests include hydride materials for solid-state hydrogen storage and semiconducting oxide nanostructures for solar energy conversion.

Preface to "Advances in Nanoparticles"

Recent advances in the synthesis of nanoparticles and in atomic-scale characterization, coupled with insights from theoretical modelling, have opened up exciting possibilities for synthesizing knowledge-based nanoparticles for many applications, such as catalysis, plasmonics, photonics, magnetism, and nanomedicine.

The number of scientific papers with "nanoparticle" as a keyword has increased almost linearly in the last 10 years from about 16,000 in 2009 to about 50,000 in 2019. This impressive worldwide interest stems from the striking scientific appeal of nanoparticles, which constitute a bridge between the atomic and bulk worlds, as well as from their actual or potential applications in many different fields. The preparation of nanoparticles is a crossroad of materials science where chemists, physicists, engineers, and biologists frequently meet, leading to a continuous improvement of existing techniques and to the invention of new methods.

The papers published in this Special Issue illustrate the fascinating richness of properties and applications that are attainable by tailoring the size, morphology, and composition of nanoparticles. For instance, concerning nanoparticle synthesis, some authors use physical methods such as pulsed laser deposition and gas phase condensation, some employ chemical methods like hydrothermal synthesis and sol-gel routes, and others even use biosynthesis. Innovation in preparation routes has produced significant benefits, for instance the one-step synthesis of nanocomposites or quantum dots. The application fields described in this Special Issue include magnetic information storage, photothermal therapy, catalysis, photonics, dye adsorption, and nanomedicine.

The reader interested in nanoparticles synthesis and properties will here find a valuable selection of scientific cases that cannot cover all methods and applications relevant to the field, but provide an updated overview on the fervent research activity focused on nanoparticles.

Luca Pasquini
Special Issue Editor

Article

One-Step Hydrothermal Synthesis of Yellow and Green Emitting Silicon Quantum Dots with Synergistic Effect

Zhixia Zhang [1,2], Chunjin Wei [1,2], Wenting Ma [1,2], Jun Li [1,2], Xincai Xiao [1,2] and Dan Zhao [1,2,*]

1. School of Pharmaceutical Sciences, South-Central University for Nationalities, Wuhan 430074, China; zhixia_z@163.com (Z.Z.); wcj407704@163.com (C.W.); tingwm1993@163.com (W.M.); lijun-pharm@hotmail.com (J.L.); xcxiao@126.com (X.X.)
2. National Demonstration Center for Experimental Ethnopharmacology Education (South-Central University for Nationalities), Wuhan 430065, China
* Correspondence: wqzhdpai@163.com; Tel.: +86-1806-208-4690

Received: 19 January 2019; Accepted: 11 March 2019; Published: 20 March 2019

Abstract: The concept of synergistic effects has been widely applied in many scientific fields such as in biomedical science and material chemistry, and has further attracted interest in the fields of both synthesis and application of nanomaterials. In this paper, we report the synthesis of long-wavelength emitting silicon quantum dots based on a one-step hydrothermal route with catechol (CC) and sodium citrate (Na-citrate) as a reducing agent pair, and N-[3-(trimethoxysilyl)propyl]ethylenediamine (DAMO) as silicon source. By controlling the reaction time, yellow-emitting silicon quantum dots and green-emitting silicon quantum dots were synthesized with quantum yields (QYs) of 29.4% and 38.3% respectively. The as-prepared silicon quantum dots were characterized by fluorescence (PL) spectrum, UV–visible spectrum, high resolution transmission electron microscope (HRTEM), Fourier transform infrared (FT-IR) spectrometry energy dispersive spectroscopy (EDS), and Zeta potential. With the aid of these methods, this paper further discussed how the optical performance and surface characteristics of the prepared quantum dots (QDs) influence the fluorescence mechanism. Meanwhile, the cell toxicity of the silicon quantum dots was tested by the 3-(4,5-dimethylthiazolyl-2)-2,5-diphenyltetrazolium (MTT) bromide method, and its potential as a fluorescence ink explored. The silicon quantum dots exhibit a red-shift phenomenon in their fluorescence peak due to the participation of the carbonyl group during the synthesis. The high-efficiency and stable photoluminescence of the long-wavelength emitting silicon quantum dots prepared through a synergistic effect is of great value in their future application as novel optical materials in bioimaging, LED, and materials detection.

Keywords: silicon quantum dots; synthesis; one-pot hydrothermal method; synergistic effect

---

1. Introduction

As a newcomer to nanomaterials, silicon quantum dots (SiQDs) have recently attracted tremendous attention. SiQDs exhibit tunable fluorescence emission properties, chemical stability, favorable biocompatibility, and low toxicity. Thus, they have shown promising potential in a wide range of fields, including lithium-ion batteries [1,2], biological imaging [3,4], and therapy [5,6].

At present, most prepared SiQDs are blue-emitting ones (λ_{em} < 450 nm) [7,8], but the short excitation wavelength (λ_{ex}) greatly limits their application in biochemical detection and imaging because of the background fluorescence interference. Therefore, the synthesis of long wavelength SiQDs has become a focus of researchers. Currently, methods for long wavelength emitting SiQDs synthesis are limited, among which the electrochemical method is commonly used. For example, Tu et al. [9] prepared red-emitting SiQDs by etching a p-type silicon wafer in an electrolyte containing HF and methanol. Kang et al. [10] synthesized SiQDs with an emission range from blue to red through a refluxing route after electrolyzing the silicon wafer in an electrolytic cell. Erogbogbo et al. [11] by pyrolyzing silane and etching, finally acquired red, yellow and green-emitting SiQDs. Besides, Kauzlarich et al. [12] acquired orange-emitting SiQDs by reflexing Mg_2S and adding normal-butyl blocking through a wet chemical method. However, these reported methods are quite complicated with obvious drawbacks such as expensive experimental equipment and high energy costs. Therefore, the development of new, simple, low energy cost, and environment-friendly synthesis methods have aroused immense interest. As a "bottom-to-up" synthesis method for SiQDs, the one-step hydrothermal route is simple in operation, and SiQDs prepared by this method have excellent dispersity without any requirement for further modification before their practical application. So far, most reported SiQDs synthesized though the one-step hydrothermal route emit blue light [7,8,13], and only a few teams have synthesized green-emitting SiQDs by selecting proper reducing agents [14–16]. In a recent work, Ma et al. [14] obtained green-emitting SiQDs with (3-Aminopropyl)triethoxysilane (APTES) as silicon source and ascorbic acid as reducing reagent (QYs = 8.2%). Wang et al. [15] acquired green-emitting SiQDs by choosing APTES as silicon source and sodium ascorbate as reducing reagent (QYs = 21%). Han et al. [16], by using DAMO as silicon source and CC as reducing reagent, finally prepared green-emitting SiQDs with QYs at 7.1%. However, the requirement for a large amount of raw materials, the low QYs, and the weak stability of prepared SiQDs turn out to be drawbacks of these syntheses. To overcome the disadvantages, synergistic effects have been introduced to improve the one-step hydrothermal route, to acquire SiQDs with better optical properties, chemical stability and high QYs.

The synergistic effect has been widely applied in the fields of material synthesis [17,18], pharmacology [19,20], and chemistry [21,22], as well as in chemical synthesis, and has been proven to effectively improve the output of the products and simplify the synthesis process. For example, Chen et al. [23] employed the synergistic effect between the metal–organic framework Pd@MOF and metal nanoparticles PdNPs to realize a one-step multiple cascade reaction to synthesize secondary arylamines; Haddleton [24] and his team used a congregation of photo-radicals based on the synergistic effect between $CuBr_2$ and tertiary amine, to speed up the synthesis of acrylate and improve the monomer conversion rate, and acquired polyacrylate with excellent uniformity and stability. Moreover, the synergistic effect also shows its effectiveness in the synthesis of fluorescence nanomaterial. The synergistic effect between different elements during the synthesis of N/S [25] or N/P [26] co-doped carbon dots could enhance the optical-chemical activity of prepared carbon dots (CDs). Recently, our team [27] reported a synergistic effect synthesis strategy for preparing SiQDs with super-high QYs (84.92%) by choosing Na-citrate and thiourea as reagent pair. Under optimal excitation environment, the prepared SiQDs emit blue fluorescence (λ_{em} = 452 nm). Therefore, the impact of the synergistic effect of double reducing reagents upon the preparation of long-wavelength-emitting QDs was proven to be effective and positive.

In this study, we report a facile one-step route to prepare stable long-wavelength-emitting SiQDs via the synergistic effect between the reducing reagents CC and Na-citrate. The impacts of different synthesis environment parameters upon the optical properties of prepared SiQDs were explored. It was discovered that the reaction temperature could effectively adjust the emission wavelength (λ_{em}) of the product. Besides, we compared the SiQDs synthesized with the reagent pair (CC and Na-citrate) and SiQDs synthesized with only Na-citrate as reagent. The size, surface properties, and elemental composition of these two SiQDs were investigated to reveal their optical characteristics and the emission mechanism. The synergistic effect between the double reducing reagents was proven to be effective in the synthesis of stable long-wavelength-emitting SiQDs with efficient photoluminescence, and is of certain significance in the application as an ideal optical material.

2. Materials and Methods

2.1. Reagents and Instruments

Sodium citrate (99.0%) was obtained from Shanghai Zhan Yun Chemical Co., Ltd. (Shanghai, China). N-[3-(Trimethoxysilyl)propyl]ethylenediamine (95.0%) and catechol ($\geq$99.0%) were purchased from Aladdin Chemistry Co., Ltd., (Shanghai, China) Sodium sulfite ($\geq$99.0%), ascorbic acid ($\geq$99.0%), thiourea ($\geq$99.0%), urea ($\geq$99.0%), dimethyl sulfoxide (DMSO) ($\geq$99.5%) and acetonitrile ($\geq$99.0%) were obtained from Sinopharm Chemical Reagent Co., Ltd. (Shanghai, China). MTT ($\geq$98.0%) was purchansed form Sigma-Aldrich Co., Ltd. (Merck KGaA, Darmstadt, Germany). All solutions were prepared using Milli-Q water (Millipore, Burlington, MA, USA) as the solvent.

UV–visible absorption spectra were acquired with a Lambda-35 UV-visible spectrophotometer (PerkinElmer Company, Waltham, MA, USA) to determine the bandgap absorption of SiQDs. Fluorescence spectra were recorded on a LS55 spectrofluorometer (PerkinElmer Company). HRTEM images were obtained with a JEM2100F transmission electron microscope (Japan Electron Optics Laboratory Company, Tokyo, Japan) and EDS data were obtained by its annex. FT-IR spectra were obtained on a Nicolet 6700 spectrometer (Thermo Fisher Scientific, Waltham, MA, USA). Zeta-potential measurement was carried out on a Zetasizer nanoseries ZEN3690 (Malvern, UK). The relative QYs of as-prepared SiQDs were measured according to the literature with Rhodamine 6G in ethanol (QY = 95%) as a reference standard. All optical measurements were performed at room temperature under ambient conditions.

2.2. Preparation of Silicon QDs

Na-citrate (9.24 $\times$ 10^{-5} mol·L^{-1}) was dissolved in water (10 mL) under stirring for 10 min. Then, DAMO (0.2 mL) was added to the solution and stirring continued for 10 min. Finally, 1.1 mg CC was added to the solution with stirring for 1 min. Subsequently, the resulting products were heated in a Teflon-equipped stainless-steel autoclave at 130 °C or 150 °C for 5 h. After naturally cooling to room temperature, four volumes of acetonitrile were added to the obtained solution and the mixture was centrifuged at 8000 rpm for 15 min to remove raw material.

The same concentration of sodium sulfite, ascorbic acid, thiourea, or urea was used instead of Na-citrate, and heated at 130 °C for 5 h to obtain SiQDs of different reducing agent combinations to compare their relative quantum yields. The relative PLQYs of as-prepared SiQDs were measured according to the literature with Rhodamine 6G in ethanol (QY = 95%) as a reference standard.

2.3. MTT Method

The MTT method was used to detect the cell viability of two kinds of SiQDs. The L02 cells were seeded at a density of 1 $\times$ 10^4 cells per well with 100 µL of culture medium in 96-well plates and placed for cell growth for 24 h in a 37 °C, 5% CO$_2$ humidified incubator. The appropriate amount of yellow-emitting SiQDs (y-SiQDs) and green-emitting SiQDs (g-SiQDs) were dissolved in ultrapure water and added to Dulbecco's modified eagle medium (DMEM) to a concentration of 200 µg/mL. Different concentrations of dilution samples were added per well (The concentration of DMSO was less than 1‰). After adding 100 µL of MTT solution to each well, it was placed in the incubator for 30 min. Its supernatant was discarded after culturing the cells for 24 h, and then 150 µL DMSO per well was added. After shaking in the dark for 10 min, the microplate reader detected the optical density value (OD) at 562 nm. The cell survival rate is calculated as follows. IC50 values were calculated by GraphPad Prism 6 (GraphPad Software, Inc., San Diego, CA, USA).

$$\text{Cell viability}(\%) = \frac{\text{OD value of experiment group} - \text{OD value of control group}}{\text{OD value of negative control group} - \text{OD value of control group}}$$

3. Results and Discussion

3.1. The Optimization of the Synthesis Environment for SiQDs

3.1.1. Filtration of the Proper Selection of Reducing Reagents for SiQDs Preparation

The reducing reagent is one of the most important factors of the synthesis that determines the optical properties of the prepared SiQDs. Proper selection of reducing reagents can effectively change the fluorescence intensity and λ_{em} of the products. Based on the published literature and our previous work, it was discovered that ascorbic acid (VitC) [14], sodium ascorbate (VitC-Na) [15], and CC [16,27] are the ideal candidates for the preparation of long-wavelength emitting SiQDs. The preparation process with one of them as reducing reagent could produce green-emitting SiQDs (λ_{em} ~ 520 nm), but the QYs of the production were quite low (<5%). Adjusting the traditional synthesis environment, such as the adjustment of reactant ratio, reaction time, and temperature, could not further improve the optical properties of the product. Therefore, this paper reports the introduction of a second reducing reagent into the synthesis process, to improve the stability and QYs of the products through a synergistic effect.

Since the synergistic effect between double reducing reagents has been proven to be effective in improving the optical properties of blue-emitting SiQDs [27], the basic reducing reagent should first be selected out of the above-mentioned three candidates (VitC, VitC-Na, and CC). Based on that, the second reducing reagent should be selected to combine with the basic reagent to improve the optical and chemical properties of the SiQDs. As is known, Na-citrate is an excellent reducing reagent [28]. The prepared SiQDs exhibit QYs as high as 73.3% with DAMO as silicon source and Na-citrate as reducing reagent. Na-citrate was thus chosen as one candidate of the reducing reagent pair. Then, three other reducing reagent candidates were matched with Na-citrate to test their ability for improvement. With the reactant ratio fixed at n(DAMO):n(CC):n(Na-Citrate) = 1:0.11:0.47 (cDAMO = 8.29×10^{-2} mol·L^{-1}), reaction temperature at 130 °C, and reaction time for 5 h, the prepared SiQDs emitted yellow fluorescence with Na-citrate/CC as reducing reagent pair. They also maintained excellent optical properties and chemical stability with QYs as high as 29.4%. Na-citrate/VitC and Na-citrate/VitC-Na pairs could be used to prepare green-emitting SiQDs (λ_{em} ~ 500 nm) with QYs of 5.5% and 6.4%. Comparing with the results of the three groups, CC was selected as one of the reducing reagents for the synergistic effect.

With CC chosen as one candidate, the other reducing reagent of the pair was changed to assess more pairing possibilities. In the synthesis process, the inorganic reducing reagent sodium sulfite, organic reducing reagents thiourea, urea, and VitC have all been proven to have excellent reducing abilities. As shown in Figure 1, the combinations of CC and any one of these reducing reagents also resulted in acquiring long-wavelength emitting SiQDs, with their QYs all higher than 15%, obviously higher than that of the SiQDs prepared by the single reducing reagent CC. This proves that the synergistic effect of double reducing reagents could effectively improve the optical properties of SiQDs. This might be attributable to the strong reducing abilities of Na-citrate, as well as the unique structure of CC which is beneficial for the red-shift of λ_{em} of the products. More detailed discussion on the synthesis mechanism is presented in the later part of this paper.

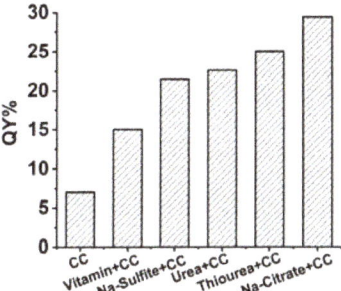

Figure 1. QYs of SiQDs synthesized with single reductant (CC) and double reductants (CC with other reductants).

3.1.2. The Impact of Synthesis Parameters upon the Optical Properties of SiQDs

The ratio of reactants, reaction temperature, and time are the key factors for SiQDs synthesis. Figure 2a shows the impact of the addition mass of Na-citrate. Fixing the reaction materials ratio of c(DAMO):c(CC) at 1:0.11, the solution was heated at 130 °C for 5 h. When the addition mass of Na-citrate was raised to 13.6 mg, the fluorescence intensity of the prepared SiQDs reached a maximum. However, the changed amounts of Na-citrate did not change the λ_{em} of the prepared SiQDs. Changing the addition amount of CC, on the other hand, had an impact upon the λ_{em}. As shown in Figure 2b, with other synthesis parameters fixed, and n(DAMO):n(Na-Citrate) set at 1:0.47, the λ_{em} of SiQDs redshifts from 518 nm to 546 nm with increasing CC amount. The QYs of the prepared SiQDs reach their maximum at 29.4% with the amount of CC rising to 1.1 mg.

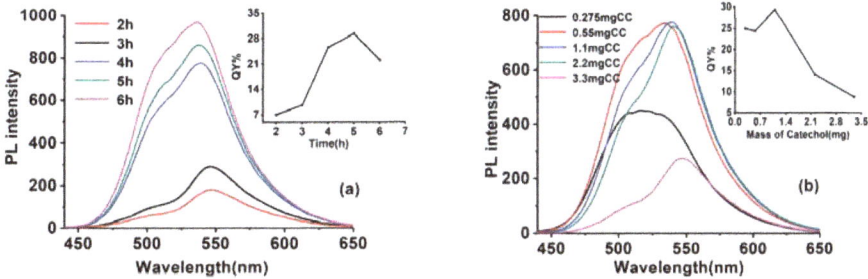

Figure 2. PL spectra of SiQDs synthesized with CC and Na-Citrate (a) with different addition mass of Na-citrate (b) with different addition masses of CC. The insets show the corresponding QYs.

On fixing the other synthesis parameters, the increased reaction time from 2 h to 6 h enhanced the fluorescence intensity of the prepared SiQDs, with the λ_{em} blue shifts from 546 nm to 535 nm (Figure 3a). The QYs of SiQDs reach their maximum with a reaction time of 5 h.

Reaction temperature is also one of the important factors affecting the optical properties of SiQDs. Figure 3b displays the different regular patterns of the QDs syntheses: the drop of reaction temperature from 150 °C to 110 °C renders λ_{em} red shifts from 520 nm to 545 nm. A significant change in the color of SiQDs from green to orange could be observed under UV light. Surface chemistry is the key factor that determines the λ_{em} change of the prepared SiQDs [29]. The long-time exposure in the high-temperature oxidation environment would destroy the structure of CC, preventing it from forming luminophores on the surface of the SiQDs while the silane mostly combines with Na-citrate, leading to short-wavelength emitting SiQDs. On the other hand, the low temperature environment would greatly decrease the fluorescence intensity. The QYs of prepared SiQDs dropped from 39.3%

to 9.4% when the reaction temperature dropped from 150 °C to 110 °C. That may because the low temperature is not beneficial to the decomposition of silane to prevent the surface defects of the SiQDs.

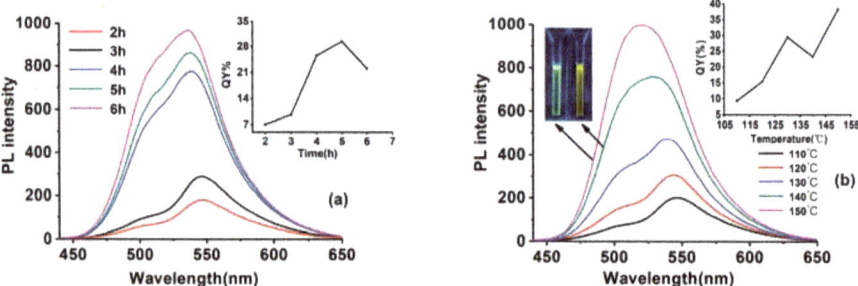

Figure 3. PL spectra of SiQDs synthesized with CC and Na-citrate at (**a**) different reaction time and (**b**) at different reaction temperature. Inset: the picture shows the increasing trend of QYs and the photos are of g-SiQDs and y-SiQDs.

Therefore, through a series of experiments, the best synthesis environment for high QYs yellow-emitting SiQDs (y-SiQDs) and green-emitting SiQDs (g-SiQDs) were acquired. As shown in the insert of Figure 3b, with n(DAMO):n(CC):n(Na-Citrate) fixed at 1:0.11:0.47, reaction time at 5 h, y-SiQDs were acquired at a temperature of 130 °C with QYs at 29.4%, and g-SiQDs were acquired at 150 °C with QYs at 38.3%. Though both were prepared with the same raw materials, these two SiQDs exhibited different optical properties. The fluorescence emission peak of y-SiQDs (Figure S1) is obviously asymmetric, and two obvious UV absorption peaks can be observed at 233 and 254 nm, while the g-SiQDs (Figure S2) does not exhibit a similar spectra shape.

3.2. Characterization of SiQDs and Mechanism Discussion

3.2.1. Three Dimensional Fluorescence Spectra

The three-dimensional fluorescence spectra could directly describe the changes of λ_{em} and fluorescence intensity with the λ_{ex}. It was thus used to compare the optical property differences of the b-SiQDs (DAMO as silicon source, Na-citrate as reducing agent), SiQDs(CC) (DAMO as silicon source, CC as reducing agent) y-SiQDs, and g-SiQDs.

Figure 4a shows the change of the b-SiQDs fluorescence emission spectra with λ_{ex} in the range of 300–420 nm (gap at 10 nm). With the increase of λ_{ex}, the fluorescence intensity continues to increase and reaches a maximum at λ_{ex} = 370 nm. Further increase of λ_{ex} leads to a rapid decrease of fluorescence intensity, with its λ_{em} staying at about 453 nm.

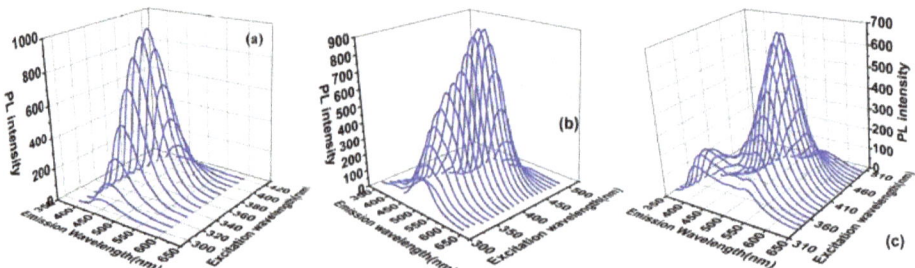

Figure 4. Three-dimensional fluorescence spectra of the three kinds of (**a**) b-SiQDs; (**b**) g-SiQDs, and (**c**) y-SiQDs.

Figure S3 shows the three-dimensional fluorescence spectra comparing the three-dimensional spectra of SiQDs(CC). We can find that, the λ_{em} of SiQDs(CC) only changes in fluorescence intensity at 550 nm with the λ_{ex} of SiQDs(CC) increasing from 310 nm to 500 nm. Also there is no occurrence of emission peak shift, which is similar to the cas of the three-dimensional spectra of b-SiQDs.

Figure 4b exhibits a change of g-SiQDs fluorescence emission spectra with λ_{ex} in the range of 310–500 nm (gap at 10 nm). The emission peak of g-SiQDs exhibits slight blue-shift when λ_{ex} increases in the range of 310–360 nm, with gradual increase of fluorescence intensity. Further increase of λ_{ex} (360–420 nm) makes the emission peak redshift with further increase of fluorescence intensity, while the fluorescence intensity reaches the maximum (λ_{em} = 527 nm) at λ_{ex} = 420 nm. When λ_{ex} reaches the range of 430–500 nm, the redshifts of the emission peak continue with decrease of fluorescence intensity.

However, the change of y-SiQDs in their three-dimensional fluorescence spectra is more complicated than the other two. Figure 4c displays the change of the y-SiQDs fluorescence emission spectra with λ_{ex} in the range of 330–550 nm (gap at 10 nm), with two different changes in behavior being observed. When λ_{ex} is in the range of 330–370 nm, two emission peaks exist at 395 nm and 540 nm. With the increase of λ_{ex}, the fluorescence intensity at 395 nm decreases, while the one at 540 nm enhances. When λ_{ex} rises up to the range of 380 to 500 nm, the emission peak at 395 nm disappears, and the fluorescence intensity of y-SiQDs increases with λ_{em}. The intensity reaches a maximum at λ_{ex} = 420 nm. Further increase of λ_{em} would lead to redshift of the emission peak and decrease of fluorescence intensity.

Therefore, the particle size distribution results of y-SiQDs were used to demonstrate the possibility of the presence of two sizes of nanoparticles in y-SiQDs. As shown in Figure S4a, when performing the particle size distribution statistics of y-SiQDs, the number of particles around ~5 nm is also higher than the normal level, in addition to the large particles of y-SiQDs. This shows that the complex behavior of the 3D spectra of y-SiQDs is likely to originate from the presence of two types of SiQDs in the sample. This phenomenon may be due to a competitive reaction between the two reducing agents and the silicon source during the synthesis. Na-citrate is an excellent reducing agent that is more easily combined with silane to produce blue-emitting SiQDs.

3.2.2. HRTEM Imaging

The particle size of SiQDs is one of the crucial factors for its optical properties. Thus, the transmission electron microscope was applied to study the particle size and morphology of the prepared b-SiQDs and y-SiQDs. Both SiQDs exhibited uniform distribution and spherical morphology (Figure 5). The diameter distribution of the b-SiQDs is in the range of 2–3.5 nm with average diameter at 2.8 nm; while the diameter distribution of y-SiQDs is in the range of 10–13 nm with average diameter at 12 nm. Figures S4 and S5 shows the particle size distribution histograms and the lattice plane of two kinds of SiQDs corresponds to the d-spacing of the cubic diamond structure of silicon giving the (111) plane with 0.31 nm spacing. It is thus proven that the increase of particle size could effectively make the emission peak redshift.

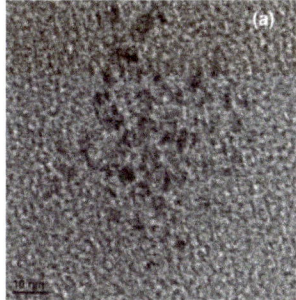 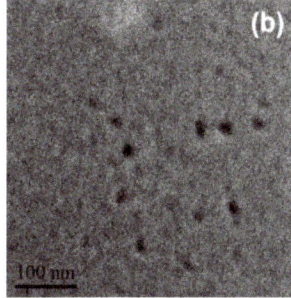

Figure 5. The TEM images of (a) b-SiQDs; (b) y-SiQDs.

3.2.3. FT-IR Spectrometer

A FT-IR spectrometer was used to examine the functional groups on the surface of SiQDs. As shown in Figure 6, the surface functional groups of b-SiQDs, G-SiQDs, and y-SiQDs are quite similar, including the stretching vibration of N–H [30], O–H [31] at 3277 cm^{-1} and 3367 cm^{-1}, and the stretching vibrations of –CH=N– [32] at 1595 cm^{-1}. Stretching and bending vibrations due to methyl groups [33] were represented by the bands at 2935 cm^{-1} and 1462 cm^{-1}, the asymmetrical deformation vibration of Si–O and the stretching vibrations of Si–O–Si at 1310 cm^{-1} and 1130–1010 cm^{-1} [27]. These results show that the existence of hydroxyl and ammonia groups could effectively enhance the water-solubility and stability of the prepared SiQDs. However, the spectra of b-SiQDs do not exhibit carbonyl groups [34] absorption peak at 1663 cm^{-1} as is the case with both y-SiQDs and g-SiQDs. The low vibration frequency of the carbonyl groups proves the presence of large conjugation systems on both sides of the carbonyl groups, leading to a weakened strength of the double bond. The enlarged conjugation system makes the λ_{em} redshift. The zeta potential of y-SiQDs is −17 mV, showing the surface of the SiQDs is negatively charged with the existence of hydroxyl group on the surface.

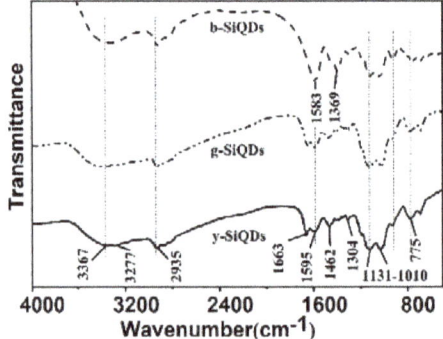

Figure 6. FT-IR spectrum of the three kinds of SiQDs (b-SiQDs, g-SiQDs, and y-SiQDs).

3.2.4. EDS Spectrum

The EDS spectrum was used to analyze the chemical constitutions of y-SiQDs and b-SiQDs (Figures S6 and S7). The contents of C, N, O, and Si elements in both SiQDs were measured, and their relative atomic weight ratios are listed in Figure 7. The relative atomic weight ratios of C to O of b-SiQDs and y-SiQDs are 1:0.05 and 1:0.168, respectively, showing a richer existence of O elements on the surface of y-SiQDs and thus a relative higher oxidation degree. This also proves the existence of carbonyl groups on the surface as shown from the FT-IR spectra of the y-SiQDs.

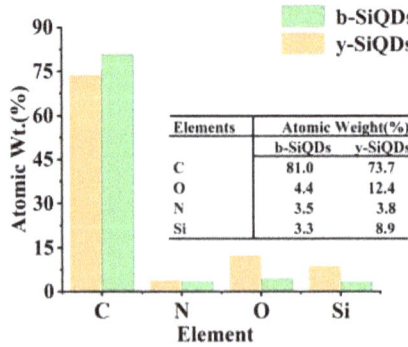

Figure 7. Atomic weight composition (%) of C, N, O, and Si for b-SiQDs and y-SiQDs examined by EDS.

3.2.5. Study on the Synthesis Mechanism

In our previous work [27], the QYs of SiQDs (with DAMO as the silicon source, sodium oxalate, and citric acid as the reducing agent pair) were effectively improved by synergistic effects, but the λ_{em} did not change. In this paper, we selected the right pair of reducing agents (CC and Na-citrate), which not only improved the QYs of SiQDs, but also caused different degrees of red shift in the emission wavelengths.

Figure 8 exhibits the synthesis mechanism of b-SiQDs, g-SiQDs, and y-SiQDs. When we use DAMO as the silicon source and Na-citrate as the reducing agent, the silane DAMO undergoes hydrolysis–reduction–polymerization to form b-SiQDs [13]; when we use DAMO as the silicon source, Na-citrate and CC as the double reducing agent, y-SiQDs are synthesized. As a common excellent reducing agent, Na-citrate is beneficial for the rapid hydrolysis and reduction of silane to form a silicon core. In addition to participating in the reduction of silane during the synthesis, the phenolic hydroxyl group on the surface of the CC can be further combined with the residue on the surface of the hydrolyzed product of DAMO to form a fluorophore having a larger conjugated system and a C=O structure on the surface, which is advantageous for the SiQDs to form long wavelengths and higher QYs. The presence of carbonyl groups on the surface of SiQDs was also confirmed by the higher O/C ratio of EDS observed in the IR spectra.

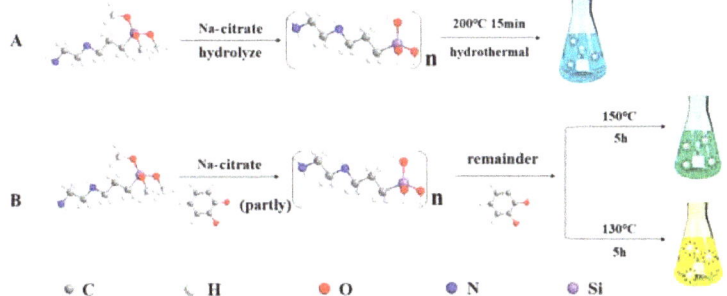

Figure 8. Schematic illustration of the synthesis of (**A**) b-SiQDs and (**B**) g-SiQDs and y-SiQDs.

It is worth noticing that lowering of either reaction temperature or time would make the emission wavelength of the prepared SiQDs redshift. This is different from the synthesis mechanism of other SiQDs. With the same reactants ratio and reaction time (5 h), y-SiQDs can be acquired at a temperature of 130 °C, while higher temperature (150 °C) would lead to the formation of g-SiQDs. We suspect that the possible reason is that this is not conducive to the stable existence of the CC structure at high temperature conditions, and thus cannot participate in the formation of the surface luminescent groups of y-SiQDs.

3.3. Application of the Prepared SiQDs

3.3.1. Optical Stability of the SiQDs

The optical stability is one of the most crucial properties of QDs for their wide application in the fields of analytical detection, optical sensing, and nonlinear optical materials. The optical stability of prepared SiQDs was tested by setting it under natural light and measuring its fluorescence intensity every five minutes. The fluorescence intensities of both y-SiQDs and g-SiQDs exhibit no obvious decrease on 30 min of exposure to natural light (Figure S8), proving their excellent optical stability.

3.3.2. Cytotoxicity of the SiQDs

The cytotoxicity of SiQDs is also an important factor with regard to their final application in an organism [35,36]. The MTT method is the common way to assess the cytotoxicity of QDs. y-SiQDs, and g-SiQDs in different concentrations were used to incubate the L02 cell for 24 h, and then the activity of the cells was measured. Figure 9 shows that two kinds of SiQDs exhibit really low cytotoxicity with almost no interference to the growth of the cells. The excellent stability and safety could guarantee their potential application as probes in biochemical fields.

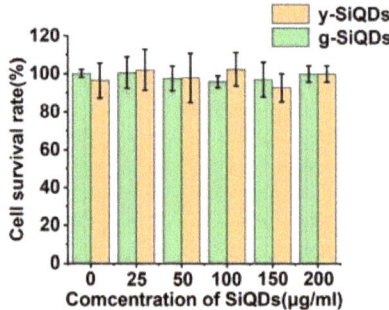

Figure 9. Cell viability (%) measured by MTT assay. The L02 cells were incubated with the SiQDs for 24 h at 37 °C. All results were presented as the mean ± standard deviation (SD) from three independent experiments with four wells in each.

3.3.3. Fluorescence Ink

Since y-SiQDs and g-SiQDs exhibit excellent optical stability and low toxicity at high concentration, they show great possibility as new biocompatible fluorescence inks. Figure 10a,b shows the pattern drawn with y-SiQDs and g-SiQDs on filter paper under natural light and UV light (365 nm). The filter paper exhibits a strong bright yellow and green fluorescence pattern under UV light. This proves that y-SiQDs and g-SiQDs could be used as a fluorescence invisible ink, or further be applied in the field of banknote anti-counterfeiting techniques [37].

Figure 10. Photos of pictures painted by y-SiQDs and g-SiQDs. Photo (**a**) was photographed under daylight and (**b**) was photographed under UV-light (365 nm).

4. Conclusions

This paper reports the preparation of stable y-SiQDs and G-SiQDs through the one-step hydrothermal route with synergistic effects between double reducing reagents CC and Na-citrate, with the silane coupling agent N-[3-(trimethoxysilyl)propyl]ethylenediamine (DAMO) as silicon source. The QYs of the prepared y-SiQDs and g-SiQDs reach as high as 29.4% and 38.3%. The optical properties of both SiQDs were characterized and tested by three-dimensional fluorescence spectra, IR spectra, EDS, and TEM, with further comparison to the b-SiQDs prepared by DAMO and Na-citrate. It was discovered that the participation of the carbonyl group during the synthesis makes the λ_{em} redshift. The preparation of long-wavelength emitting SiQDs through a synergistic effect and their highly efficient and stable photoluminescence ensure their promising application as optical materials.

Supplementary Materials: The following are available online at http://www.mdpi.com/2079-4991/9/3/466/s1, Figure S1: UV–Vis adsorption and photoluminescence emission spectra of y-SiQDs; Figure S2: UV–Vis adsorption and photoluminescence emission spectra of g-SiQDs; Figure S3: Three-dimensional fluorescence spectra of three kinds of SiQDs(CC); Figure S4: (a) particle size distribution histograms and (b) lattice image y-SiQDs; Figure S5: (a) particle size distribution histograms and (b) lattice image b-SiQDs; Figure S6: EDS spectra of y-SiQDs; Figure S7: EDS spectra of b-SiQDs; Figure S8: The relationship of replaced time with the fluorescence intensity of y-SiQDs and g-SiQDs.

Author Contributions: Conceptualization, D.Z. and Z.Z.; methodology, D.Z. and Z.Z.; software, J.L.; project administration, D.Z., J.L., and X.X.; formal analysis, Z.Z.; investigation, Z.Z., C.W., and W.M.; resources, D.Z.; data curation, Z.Z.; writing—original draft preparation, Z.Z.; writing—review and editing, Z.Z. and D.Z.; visualization, Z.Z., C.W., and W.M.; supervision, D.Z.; funding acquisition, D.Z.

Funding: This research was funded by NATURAL SCIENCE FOUNDATION OF HUBEI PROVINCE, grant number 2016CFB615 and by FUNDAMENTAL RESEARCH FUNDS FOR THE "CENTRAL UNIVERSITIES", SOUTH CENTRAL UNIVERSITY FOR NATIONALITIES, grant number CZY19029, CZY15020, CZW15017 and XTZ15013.

Acknowledgments: The authors would like to thank Jianhua Wei for his technical support.

Conflicts of Interest: The authors declare no conflict of interest.

References

1. Shan, C.; Wu, K.; Yen, H.-J.; Narvaez Villarrubia, C.; Nakotte, T.; Bo, X.; Zhou, M.; Wu, G.; Wang, H.-L. Graphene Oxides Used as a New "Dual Role" Binder for Stabilizing Silicon Nanoparticles in Lithium-Ion Battery. *ACS Appl. Mater. Interfaces* **2018**, *10*, 15665–15672. [CrossRef] [PubMed]
2. Tang, W.; Guo, X.X.; Liu, X.H.; Chen, G.; Wang, H.J.; Zhang, N.; Wang, J.; Qiu, G.Z.; Ma, R.Z. Interconnected silicon nanoparticles originated from halloysite nanotubes through the magnesiothermic reduction: A high-performance anode material for lithium-ion batteries. *Appl. Clay Sci.* **2018**, *162*, 499–506. [CrossRef]
3. Tang, M.M.; Ji, X.Y.; Xu, H.; Zhang, L.; Jiang, A.R.; Song, B.; Su, Y.Y.; He, Y. Photostable and Biocompatible Fluorescent Silicon Nanoparticles-Based Theranostic Probes for Simultaneous Imaging and Treatment of Ocular Neovascularization. *Anal. Chem.* **2018**, *90*, 8188–8195. [CrossRef] [PubMed]
4. Geng, X.; Li, Z.H.; Hu, Y.L.; Liu, H.F.; Sun, Y.Q.; Meng, H.M.; Wang, Y.W.; Qu, L.B.; Lin, Y.H. One-Pot Green Synthesis of Ultrabright N-Doped Fluorescent Silicon Nanoparticles for Cellular Imaging by Using Ethylenediaminetetraacetic Acid Disodium Salt as an Effective Reductant. *ACS Appl. Mater. Interfaces* **2018**, *10*, 27979–27986. [CrossRef] [PubMed]
5. Wang, R.G.; Zhao, M.Y.; Deng, D.; Ye, X.; Zhang, F.; Chen, H.; Kong, J.L. An intelligent and biocompatible photosensitizer conjugated silicon quantum dots-MnO$_2$ nanosystem for fluorescence imaging-guided efficient photodynamic therapy. *J. Mater. Chem. B* **2018**, *6*, 4592–4601. [CrossRef]
6. Liu, Z.H.; Li, Y.Z.; Li, W.; Xiao, C.; Liu, D.F.; Dong, C.; Zhang, M.; Makila, E.; Kemell, M.; Salonen, J.; et al. Multifunctional Nanohybrid Based on Porous Silicon Nanoparticles, Gold Nanoparticles, and Acetalated Dextran for Liver Regeneration and Acute Liver Failure Theranostics. *Adv. Mater.* **2018**, *30*, 1703393–1703403. [CrossRef] [PubMed]
7. Luo, L.; Song, Y.; Zhu, C.Z.; Fu, S.F.; Shi, Q.R.; Sun, Y.M.; Jia, B.Z.; Du, D.; Xu, Z.L.; Lin, Y.H. Fluorescent silicon nanoparticles-based ratiometric fluorescence immunoassay for sensitive detection of ethyl carbamate in red wine. *Sens. Actuators B Chem.* **2018**, *255*, 2742–2749. [CrossRef]

8. Zhu, B.Y.; Ren, G.J.; Tang, M.Y.; Chai, F.; Qu, F.Y.; Wang, C.G.; Su, Z.M. Fluorescent silicon nanoparticles for sensing Hg^{2+} and Ag^+ as well visualization of latent fingerprints. *Dyes Pigments* **2018**, *149*, 686–695. [CrossRef]
9. Tu, C.C.; Hoo, J.H.; Bohringer, K.F.; Lin, L.Y.; Cao, G.Z. Red-emitting silicon quantum dot phosphors in warm white LEDs with excellent color rendering. *Opt. Express* **2014**, *22*, A276–A281. [CrossRef] [PubMed]
10. Kang, Z.; Liu, Y.; Tsang, C.H.A.; Ma, D.D.D.; Fan, X.; Wong, N.B.; Lee, S.T. Water-soluble silicon quantum dots with wavelength-tunable photoluminescence. *Adv. Mater.* **2009**, *21*, 661–664. [CrossRef]
11. Erogbogbo, F.; Yong, K.T.; Roy, I.; Xu, G.X.; Prasad, P.N.; Swihart, M.T. Biocompatible luminescent silicon quantum dots for imaging of cancer cells. *ACS Nano* **2008**, *2*, 873–878. [CrossRef] [PubMed]
12. Dohnalova, K.; Poddubny, A.N.; Prokofiev, A.A.; de Boer, W.D.A.M.; Umesh, C.P.; Paulusse, J.M.J.; Zuilhof, H.; Gregorkiewicz, T. Surface brightens up Si quantum dots: direct bandgap-like size-tunable emission. *Light Sci. Appl.* **2013**, *2*, 47–53. [CrossRef]
13. Xu, X.L.; Ma, S.Y.; Xiao, X.C.; Hu, Y.; Zhao, D. The preparation of high-quality water-soluble silicon quantum dots and their application in the detection of formaldehyde. *RSC Adv.* **2016**, *6*, 98899–98907. [CrossRef]
14. Ma, S.D.; Chen, Y.L.; Feng, J.; Liu, J.J.; Zuo, X.W.; Chen, X.G. One-Step Synthesis of Water-Dispersible and Biocompatible Silicon Nanoparticles for Selective Heparin Sensing and Cell Imaging. *Anal. Chem.* **2016**, *88*, 10474–10481. [CrossRef]
15. Wang, J.; Ye, D.X.; Liang, G.H.; Chang, J.; Kong, J.L.; Chen, J.Y. One-step synthesis of water-dispersible silicon nanoparticles and their use in fluorescence lifetime imaging of living cells. *J. Mater. Chem. B* **2014**, *2*, 4338–4345. [CrossRef]
16. Han, Y.X.; Chen, Y.L.; Feng, J.; Liu, J.J.; Ma, S.D.; Chen, X.G. One-Pot Synthesis of Fluorescent Silicon Nanoparticles for Sensitive and Selective Determination of 2,4,6-Trinitrophenol in Aqueous Solution. *Anal. Chem.* **2017**, *89*, 3001–3008. [CrossRef] [PubMed]
17. Srivastava, R.; Gupta, S.K.; Naaz, F.; Singh, A.; Singh, V.K.; Verma, R.; Singh, N.; Singh, R.K. Synthesis, antibacterial activity, synergistic effect, cytotoxicity, docking and molecular dynamics of benzimidazole analogues. *Comput. Biol. Chem.* **2018**, *76*, 1–16. [CrossRef] [PubMed]
18. Yang, Q.; Xu, Q.; Jiang, H.-L. Metal–organic frameworks meet metal nanoparticles: synergistic effect for enhanced catalysis. *Chem. Soc. Rev.* **2017**, *46*, 4774–4808. [CrossRef] [PubMed]
19. Liu, C.B.; Chu, X.J.; Sun, P.Y.; Feng, X.J.; Huang, W.W.; Liu, H.X.; Ma, Y.B. Synergy effects of Polyinosinic-polycytidylic acid, CpG oligodeoxynucleotide, and cationic peptides to adjuvant HPV E7 epitope vaccine through preventive and therapeutic immunization in a TC-1 grafted mouse model. *Hum. Vaccines Immunother.* **2018**, *14*, 931–940. [CrossRef]
20. Zhao, R.L.; He, Y.M. Network pharmacology analysis of the anti-cancer pharmacological mechanisms of Ganoderma lucidum extract with experimental support using Hepa1-6-bearing C57 BL/6 mice. *J. Ethnopharmacol.* **2018**, *210*, 287–295. [CrossRef]
21. Zhan, Y.Q.; Zhang, J.M.; Wan, X.Y.; Long, Z.H.; He, S.J.; He, Y. Epoxy composites coating with Fe_3O_4 decorated graphene oxide: Modified bio-inspired surface chemistry, synergistic effect and improved anti-corrosion performance. *Appl. Surf. Sci.* **2018**, *436*, 756–767. [CrossRef]
22. Zhang, L.; Zhang, G.; Li, Y.; Wang, S.; Lei, A. The synergistic effect of self-assembly and visible-light induced the oxidative C–H acylation of N-heterocyclic aromatic compounds with aldehydes. *Chem. Commun.* **2018**, *54*, 5744–5747. [CrossRef] [PubMed]
23. Chen, Y.Z.; Zhou, Y.X.; Wang, H.; Lu, J.; Uchida, T.; Xu, Q.; Yu, S.H.; Jiang, H.L. Multifunctional PdAg@MIL-101 for one-pot cascade reactions: combination of host–guest cooperation and bimetallic synergy in catalysis. *ACS Catal.* **2015**, *5*, 2062–2069. [CrossRef]
24. Anastasaki, A.; Nikolaou, V.; Zhang, Q.; Burns, J.; Samanta, S.R.; Waldron, C.; Haddleton, A.J.; McHale, R.; Fox, D.; Percec, V.; et al. Copper(II)/Tertiary Amine Synergy in Photoinduced Living Radical Polymerization: Accelerated Synthesis of omega-Functional and alpha,omega-Heterofunctional Poly(acrylates). *J. Am. Chem. Soc.* **2014**, *136*, 1141–1149. [CrossRef]
25. Ruiz-Palomero, C.; Soriano, M.L.; Benitez-Martinez, S.; Valcarcel, M. Photoluminescent sensing hydrogel platform based on the combination of nanocellulose and S,N-codoped graphene quantum dots. *Sens. Actuators B Chem.* **2017**, *245*, 946–953. [CrossRef]

26. Xiao, N.; Liu, S.G.; Mo, S.; Li, N.; Ju, Y.J.; Ling, Y.; Li, N.B.; Luo, H.Q. Highly selective detection of p-nitrophenol using fluorescence assay based on boron, nitrogen co-doped carbon dots. *Talanta* **2018**, *184*, 184–192. [CrossRef] [PubMed]
27. Ma, S.Y.; Yue, T.; Xiao, X.C.; Cheng, H.; Zhao, D. A proof of concept study of preparing ultra bright silicon quantum dots based on synergistic effect of reductants. *J. Lumin.* **2018**, *201*, 77–84. [CrossRef]
28. Wu, F.G.; Zhang, X.D.; Kai, S.Q.; Zhang, M.Y.; Wang, H.Y.; Myers, J.N.; Weng, Y.X.; Liu, P.D.; Gu, N.; Chen, Z. One-Step Synthesis of Superbright Water-Soluble Silicon Nanoparticles with Photoluminescence Quantum Yield Exceeding 80%. *Adv. Mater. Interfaces* **2015**, *2*, 1500360–1500371. [CrossRef]
29. Ghosh, B.; Shirahata, N. Colloidal silicon quantum dots: synthesis and luminescence tuning from the near-UV to the near-IR range. *Sci. Technol. Adv. Mater.* **2014**, *15*, 014207–014221. [CrossRef]
30. Kumar, A.; Grewal, A.S.; Singh, V.; Narang, R.; Pandita, D.; Lather, V. Synthesis, Antimicrobial Activity and QSAR Studies of Some New Sparfloxacin Derivatives. *Pharm. Chem. J.* **2018**, *52*, 444–454. [CrossRef]
31. Selim, Y.; Abd El-Azim, M.H.M. Conventional and Microwave-Activated the Synthesis of a Novel Series of Imidazoles, Pyrimidines, and Thiazoles Candidates. *J. Heterocycl. Chem.* **2018**, *55*, 1403–1409. [CrossRef]
32. Oleynik, I.V.; Oleynik, I.I. Design of Postmetallocene Catalytic Systems of Arylimine Type for Olefin Polymerization: XVIII. Synthesis of N-Arylsalicylaldimine Ligands Containing meta-or para-Diallylamino Group and Their Titanium(IV) Complexes. *Russ. J. Org. Chem.* **2018**, *54*, 537–544. [CrossRef]
33. Musa, W.J.; Duengo, S.; Situmeang, B. Isolation and characterization triterpenoid compound from leaves mangrove plant (Sonnertia Alba) and antibacterial activity test. *Int. Res. J. Pharm.* **2018**, *9*, 85–89. [CrossRef]
34. Kyomugasho, C.; Christiaens, S.; Shpigelman, A.; Van Loey, A.M.; Hendrickx, M.E. FT-IR spectroscopy, a reliable method for routine analysis of the degree of methylesterification of pectin in different fruit- and vegetable-based matrices. *Food Chem.* **2015**, *176*, 82–90. [CrossRef] [PubMed]
35. Liu, P.; Behray, M.; Wang, Q.; Wang, W.; Zhou, Z.G.; Chao, Y.M.; Bao, Y.P. Anti-cancer activities of allyl isothiocyanate and its conjugated silicon quantum dots. *Sci. Rep.* **2018**, *8*, 1084. [CrossRef] [PubMed]
36. Cheng, X.Y.; McVey, B.F.P.; Robinson, A.B.; Longatte, G.; O'Mara, P.B.; Tan, V.T.G.; Thordarson, P.; Tilley, R.D.; Gaus, K.; Gooding, J.J. Protease sensing using nontoxic silicon quantum dots. *J. Biomed. Opt.* **2017**, *22*, 1–7. [CrossRef] [PubMed]
37. Song, B.; Wang, H.Y.; Zhong, Y.L.; Chu, B.B.; Su, Y.Y.; He, Y. Fluorescent and magnetic anti-counterfeiting realized by biocompatible multifunctional silicon nanoshuttle-based security ink. *Nanoscale* **2018**, *10*, 1617–1621. [CrossRef] [PubMed]

© 2019 by the authors. Licensee MDPI, Basel, Switzerland. This article is an open access article distributed under the terms and conditions of the Creative Commons Attribution (CC BY) license (http://creativecommons.org/licenses/by/4.0/).

Article

One-Step Synthesis of Metal/Oxide Nanocomposites by Gas Phase Condensation

Nicola Patelli [1,*], Andrea Migliori [2], Vittorio Morandi [2] and Luca Pasquini [1,*]

[1] Department of Physics and Astronomy, Alma Mater Studiorum Università di Bologna, Viale Berti-Pichat 6/2, 40127 Bologna, Italy
[2] Section of Bologna, Institute of Microelectronics and Microsystems, National Research Council, Via Gobetti 101, 40129 Bologna, Italy; migliori@bo.imm.cnr.it (A.M.); morandi@bo.imm.cnr.it (V.M.)
* Correspondence: nicola.patelli@unibo.it (N.P.); luca.pasquini@unibo.it (L.P.)

Received: 5 January 2019; Accepted: 2 February 2019; Published: 6 February 2019

Abstract: Metallic nanoparticles (NPs), either supported on a porous oxide framework or finely dispersed within an oxide matrix, find applications in catalysis, plasmonics, nanomagnetism and energy conversion, among others. The development of synthetic routes that enable to control the morphology, chemical composition, crystal structure and mutual interaction of metallic and oxide phases is necessary in order to tailor the properties of this class of nanomaterials. With this work, we aim at developing a novel method for the synthesis of metal/oxide nanocomposites based on the assembly of NPs formed by gas phase condensation of metal vapors in a He/O_2 atmosphere. This new approach relies on the independent evaporation of two metallic precursors with strongly different oxidation enthalpies. Our goal is to show that the precursor with less negative enthalpy gives birth to metallic NPs, while the other to oxide NPs. The selected case study for this work is the synthesis of a Fe-Co/TiO_x nanocomposite, a system of great interest for its catalytic and magnetic properties. By exploiting the new concept, we achieve the desired target, i.e., a nanoscale dispersion of metallic alloy NPs within titanium oxide NPs, the structure of which can be tailored into $TiO_{1-\delta}$ or TiO_2 by controlling the synthesis and processing atmosphere. The proposed synthesis technique is versatile and scalable for the production of many NPs-assembled metal/oxide nanocomposites.

Keywords: nanoparticles; nanocomposites; gas phase condensation; electron microscopy; metal oxides; alloys; iron; cobalt; titanium

---

1. Introduction

Oxide-supported metal nanoparticles (NPs) are a class of functional materials that find innovative applications in many materials science fields such as catalysis for the production of synthetic hydrocarbons [1–3] and CO reduction [4], chemical synthesis [5,6], nanoplasmonics [7] for the development of higher efficiency photovoltaic cells [5,8], and magnetism [9]. The presence of the oxide support within the nanocomposite does not only affect the size and shape of metal NPs [10,11], but is also crucial to prevent coarsening and sintering [12], and is often responsible for a radical change in physical properties because of electronic interactions at interfacial sites [4,13,14].

In the last decade, much effort has been spent to develop novel and flexible synthesis routes for metal/oxide nanocomposites (NCs). Most of these techniques involve two-step processes, in which a porous oxide host (typically zeolites, Al_2O_3) or oxide NPs [15] are imbued with a colloidal suspension of metallic NPs produced via physical (e.g., pulsed laser ablation in liquid) or chemical methods (e.g., precipitation and nitride impregnation). Metal NP encapsulation into oxide shells [16] and in pores and channels of hierarchical zeolites has been also reported but, albeit innovative, these approaches do not convey a homogeneous distribution of the supported NPs and are limited to a small range of materials [17]. Mechanochemistry via ball milling followed by suitable thermal treatments can be successfully applied to the synthesis of metallic NPs in an oxide matrix [18]. However, it is not possible to control the morphology of the metal and oxide particles independently, and ductile materials are very difficult to process. Metallic NPs embedded in an oxide matrix can be prepared by a sol–gel method [19,20], again with some limitations on the independent control of the two phases. The deposition of metallic NPs on oxide surfaces is of great importance for fundamental studies on model systems [21], but cannot be used for the synthesis of 3D bulk materials.

In this work, we present a novel one-step strategy for the synthesis of metal/oxide NCs through the physical assembly of NPs. More specifically, NPs are formed by gas phase condensation of metallic vapors in a He/O_2 mixed atmosphere. Two metallic precursors with different oxidation enthalpies are evaporated simultaneously and independently; the one with less negative enthalpy forms metallic NPs, while the other provides the seed for oxide NPs. Thermal treatments in suitable atmosphere can be further applied to modify structure and morphology. We apply this concept to the synthesis of a Fe-Co/TiO_x NC. We also demonstrate how the stoichiometry and crystalline structure of TiO_x can be tailored by controlling the O_2 partial pressure during the synthesis and processing atmosphere. The presented method is general and scalable for production of 3D oxide-supported metal NPs.

2. Materials and Methods

Fe-Co/TiO_x nanocomposites (NCs) were grown by gas phase condensation (GPC) in an ultra-high vacuum (UHV) chamber starting from Ti (99.9%), Fe (99.9%) and Co (99.9%) powders. A schematic sketch of the system is shown in Figure 1. The main chamber is equipped with two thermal evaporation sources (Joule-heated tungsten boats). Individual evaporation rates can be monitored using a quartz crystal balance positioned close to the collection cylinder.

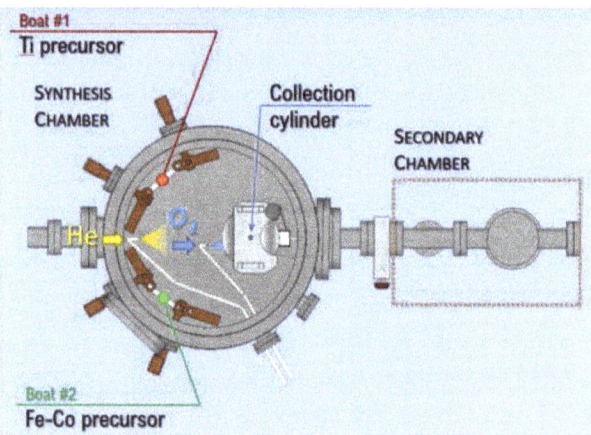

Figure 1. Top view of the GPC apparatus.

During the synthesis, He (99.9996% purity) and O_2 (99.9999% purity) are fed into the chamber (previously evacuated to 2×10^{-5} Pa) using two mass flow controllers, while the total pressure is maintained at 260 Pa by means of a rotary pump. In all the experiments reported here the O_2 content in the atmosphere was kept below 1 mol.%. The precursor materials are evaporated slightly above their melting point. NPs nucleation takes place in the gas phase where metal vapors rapidly supersaturate because of thermalization with the He/O_2 atmosphere. The NPs are collected onto the rotating stainless-steel cylinder filled with liquid N_2. Finally, the NPs are scraped off the cylinder and transferred into the secondary UHV chamber, which is equipped with an independent pumping system. Here it is possible to press the NPs into a pellet and/or to perform thermal treatments under vacuum or controlled atmosphere.

The average elemental composition of the NCs was determined using a Leica Cambridge Stereoscan 360 scanning electron microscope (SEM) equipped with Oxford Instruments X-ray detector for energy dispersive X-ray microanalysis (EDX) (Oxford Instruments, Abingdon-on-Thames, UK). X-ray diffraction (XRD) patterns were collected using a PANalytical X'celerator powder diffractometer (Malvern Panalytical, Malvern, UK) employing Cu Kα radiation (λ = 1.5406 Å). The patterns were recorded under ambient air in about 30 min. Quantitative analysis based on the Rietveld method was carried out with the MAUD program [22] to determine the lattice parameters, crystallite size and phase abundance.

Elemental mapping and phase distribution at the nanoscale were investigated with a FEI Tecnai F20 ST transmission electron microscope (TEM) (FEI Company, Hillsboro, OR, USA). EDX profiling and elemental mapping with a spatial resolution of 2 nm were recorded in scanning transmission mode (STEM) at 200 kV. The crystalline phase distribution was determined by operating in selected area diffraction (SAD) and High Resolution (HR-TEM) mode. For TEM analysis, the samples were dispersed in isopropanol, sonicated and the NPs suspension was drop-casted on a holey carbon grid.

2.1. Synthesis of NPs and NCs Samples

2.1.1. TiO_x NPs

This work can be divided in three main parts, the first regarding the influence of O_2 content in the atmosphere and of post-synthesis treatments on the stoichiometry and structure of Ti oxide NPs (from here indicated TiO_x NPs). Two samples, named Ti-O_l and Ti-O_h, were synthetized at O_2 partial pressures of 0.4 and 2.2 Pa, respectively. Table 1 lists the conditions applied during the synthesis and the successive treatments. The thermal treatments were performed either in H_2 (99.995% purity), Ar (99.999% purity) or air in a tubular stainless-steel oven at T = 400 °C for 24 h.

Table 1. Gas flow and O_2 partial pressure during the synthesis of TiO_x NPs. The total pressure was 260 Pa. The samples were examined as-prepared and after being subjected to different thermal treatments as indicated.

Sample	Inlet Flow [nmL/min]		O_2 Partial P [Pa]	Post-Synthesis Treatment (P = 0.1 MPa, T = 400 °C)		
	He	O_2		H_2	Ar	air
Ti-O_l	60.0	0.1	0.4	x	x	x
Ti-O_h	60.0	0.5	2.2			x

2.1.2. Fe-Co Alloy NPs

The second part of this work concerns the synthesis of Fe-Co alloy NPs in a He atmosphere. To this purpose, one thermal source is loaded with the desired mixture of Fe and Co powders. The powders are melted under high vacuum and rapidly cooled to about 1000 °C in order to homogenize the alloy precursor. Four samples with composition $Fe_{100-x}Co_x$ (with x = 0, 23, 48, 68 at.%) were prepared. Table 2 lists the Fe and Co content of all as-prepared NPs as determined from EDX analysis.

We notice that the NPs composition is compatible (within the uncertainties) with the precursor composition for all samples but the Co-richest one. For the Fe-Co system, the stoichiometry can be quite well preserved because Fe and Co have similar vapor pressures. Nevertheless, the slightly higher evaporation rate of Fe [23] can be the reason for the small increase in the Fe content noticeable in Table 2 for the Co-richest sample.

Table 2. Fe-Co alloy NPs. Comparison between the composition of the precursor powder mixture and of the synthesized NPs. The samples names reflect the NPs composition as determined by SEM-EDX.

Sample	Fe-Co Precursor Mixture [at%]		Fe-Co NPs [at%]	
	Fe	Co	Fe	Co
Fe$_{100}$	100	0	100	0
Fe$_{77}$Co$_{23}$	76(1)	24(1)	77(1)	23(1)
Fe$_{52}$Co$_{48}$	51(1)	49(1)	52(1)	48(1)
Fe$_{32}$Co$_{68}$	27(1)	73(1)	32(1)	68(1)

2.1.3. Fe/TiO$_x$ and Fe-Co/TiO$_x$ NCs

To obtain Fe$_{100-x}$Co$_x$ NPs (x = 0, 50) supported on TiO$_x$ NPs, the evaporation chamber was equipped with two tungsten boats (as shown in Figure 1). The boats were separated by ~30 cm in order to avoid the mixing of metal vapors before NPs nucleation [24], which may lead to the nucleation of a ternary Ti-Fe-Co alloy. The synthesis parameters are reported in Table 3. The NCs were grown in a He atmosphere with the same O$_2$ content as for the Ti-O_1 sample. The individual evaporation rates of Fe-Co and Ti were monitored with the quartz crystal balance and tuned in order to obtain a Fe-Co content in the NCs of about 10 wt%. The relative Fe/Co content in the NCs determined by SEM-EDX was consistent with the precursor within the uncertainties.

Table 3. Gas flow and O$_2$ partial pressure during the synthesis of Fe-Co/TiO$_x$ NCs. The total pressure was 260 Pa. The Co and Fe content in the mixed powder precursor are reported.

Sample	Inlet Flow [nmL/min]		O$_2$ Partial P [Pa]	Fe-Co Precursor Mixture [at%]	
	He	O$_2$		Fe	Co
Fe/TiO$_x$	60.0	0.1	0.4	100	-
Fe$_{50}$Co$_{50}$/TiO$_x$	60.0	0.1	0.4	50(1)	50(1)

3. Results and Discussion

3.1. TiO$_x$ NPs

The synthesis conditions, as well as the post-synthesis treatments, turn out to have a great influence on the structure and phase of the TiO$_x$ NPs, as summarized in Figure 2.

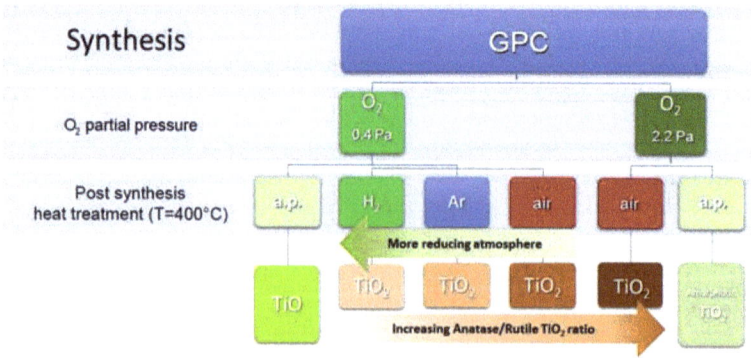

Figure 2. Schematics of the phases identified in TiO$_x$ NPs as a function of the conditions applied during synthesis by GPC and successive thermal treatments.

3.1.1. As-Prepared Samples

Figure 3 shows that the structure of the as-prepared TiO_x NPs strongly depends on the O_2 content in the atmosphere. At low O_2 (sample Ti-O_l) crystalline Ti monoxide ($TiO_{1-\delta}$) NPs are obtained (Figure 3a). $TiO_{1-\delta}$ has a disordered nonstoichiometric rocksalt structure that can exist over a wide compositional range, from $\delta = 0.30$ to $\delta = -0.25$, due to the presence of vacancies in both the Ti and O sub-lattices [25–27]. It is kinetically stable up to about 400 °C.

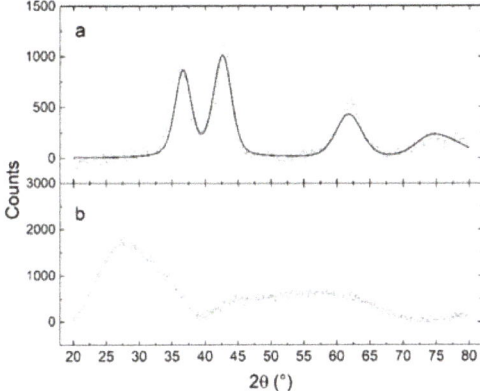

Figure 3. XRD patterns of as-prepared TiO_x NPs. (**a**) Ti-O_l; (**b**) Ti-O_h. The grey circles represent experimental data. The result of Rietveld refinement is shown for Ti-O_l as a black solid line.

The broadening of $TiO_{1-\delta}$ XRD peaks in Figure 3a is due to the small crystallite size d_{TiO} and to the high root-mean-square microstrain ε_{rms}. The relative height of the peaks is strongly influenced by the occupancy factor of Ti and O sub-lattices and permits to estimate the average stoichiometry of the NPs. From the quantitative Rietveld analysis, we obtained $d_{TiO} = 9 \pm 2$ nm, $\varepsilon_{rms} \approx 2\%$, and $\delta = 0.25 \pm 0.05$, i.e., an oxygen-deficient stoichiometry. However, the high microstrain points to a nonhomogeneous stoichiometry across the sample. In fact, the poor quality of the fit in Figure 3a indicates that a model with a single value of δ and d_{TiO} does not represent satisfactorily the structure of $TiO_{1-\delta}$ NPs. We suggest that the δ value may experience significant fluctuations among the NPs depending on their diameter and on small variations in the vapor pressure during the synthesis.

At higher O_2 content in the synthesis atmosphere (sample Ti-O_h), two broad humps indicate the formation of amorphous TiO_2 (Figure 3b) [28], while the Bragg reflections of $TiO_{1-\delta}$ are not detected.

3.1.2. Post-Synthesis Thermal Treatment

Figure 4 displays the XRD patterns of TiO_x NPs subjected to thermal treatments at T = 400 °C. In sample Ti-O_l, the Bragg reflections of $TiO_{1-\delta}$ disappear in favor of those associated with rutile and anatase TiO_2 (Figure 4a–c). The anatase content (see Table 4) increases with the oxidative power of the atmosphere ranging from 56 ± 1 wt% (in H_2) to 74 ± 2 wt% (in air). The previously amorphous sample Ti-O_h appears completely crystallized and exhibits the highest anatase content (84 ± 1 wt%). The crystallite size of anatase is not significantly influenced by the treatment conditions and varies in the 13–16 nm range. The crystallite size of rutile is smaller (6–11 nm) and appears negatively correlated with the rutile content. The HR-TEM image in Figure 4e reveals the coexistence of rutile and anatase polymorphs on a nanoscale level and confirms that no residual metallic Ti is left after the treatments, in agreement with the results of previous X-ray absorption experiments [28].

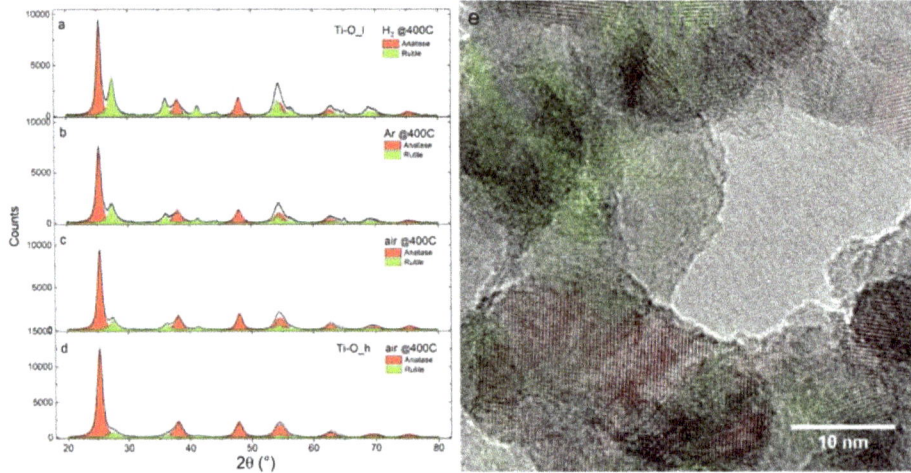

Figure 4. XRD patterns of TiO$_x$ NPs samples after a thermal treatment at 400 °C. (**a–c**) sample Ti-O_l in (**a**) H$_2$, (**b**) Ar, and (**c**) air; (**d**) sample Ti-O_h in air. The results of Rietveld refinement are reported in Table 4; (**e**) HR-TEM image of sample Ti-O_l treated in air (corresponding to pattern [c]) with the anatase and rutile fringes highlighted in red and green, respectively.

Table 4. Quantitative phase analysis for TiO$_x$ NPs samples after thermal treatments: rutile vs anatase abundance (wt%), crystallite size d and lattice parameters a and c.

Treatment Atmosphere	Anatase TiO$_2$				Rutile TiO$_2$			
	wt%	d [nm]	a [Å]	c [Å]	wt%	d [nm]	a [Å]	c [Å]
				Ti-O_l				
H$_2$	56(1)	16(2)	3.7895(3)	9.465(2)	44(1)	11(1)	4.5970(7)	2.9568(8)
Ar	64(1)	14(2)	3.790(1)	9.459(2)	36(1)	8(1)	4.600(1)	2.956(2)
air	74(2)	15(2)	3.7855(3)	9.461(2)	26(2)	8(2)	4.597(2)	2.945(2)
				Ti-O_h				
air	84(1)	13(1)	3.798(2)	9.4691(4)	16(1)	6.0(5)	4.595(6)	2.945(6)

The results related to synthesis and processing of TiO$_x$ NPs can be rationalized as follows. If the O$_2$ content in the synthesis atmosphere is too low (sample Ti-O_l), full oxidation of the nucleated Ti NPs into TiO$_2$ is not possible, and non-stoichiometric TiO$_{1-\delta}$ is obtained. This is not because the initial O$_2$ partial pressure (0.4 Pa) is below the equilibrium pressure for TiO$_2$ formation (which is ridiculously low, i.e., ~10^{-59} Pa at 400 °C and ~10^{-150} Pa at room temperature) but because there is not enough O$_2$ available. The freshly evaporated Ti consumes almost all the O$_2$ and the competition between the evaporation rate and the O$_2$ inlet flow rate dictates the final stoichiometry. Therefore, it should be possible to tailor the off-stoichiometry δ by playing with these parameters; this may be the subject of future experiments. It is also worth noticing that the TiO$_{1-\delta}$ NPs obtained in this way are kinetically stable at room temperature, i.e., they are not oxidized into TiO$_2$ upon exposure to ambient air.

Conversely, at sufficiently high O$_2$ inlet flow (sample Ti-O_h), it is possible to achieve (almost) full oxidation of the NPs. It is well known that amorphous TiO$_2$ is obtained when oxidation is carried out close to room temperature, whereas oxidation above 350 °C leads to crystalline TiO$_2$ [29]. This suggests that NPs oxidation takes place after cool down from the evaporation temperature has taken place via thermalization with the surrounding He gas. In agreement with the vast literature on TiO$_2$, crystallization of amorphous NPs is induced by a thermal treatment in air at 400 °C [28,30].

The formation of crystalline TiO$_2$ by heating the TiO$_{1-\delta}$ NPs in Ar or H$_2$ is less straightforward, but can be understood from thermodynamic data considering the presence of water vapor impurities. Let us take into account the following reaction, pertinent to the treatment in H$_2$:

$$\text{TiO} + \text{H}_2\text{O} \leftrightarrow \text{TiO}_2 + \text{H}_2 \tag{1}$$

The ratio between the water vapor pressure P$_{H2O}$ and hydrogen pressure P$_{H2}$, at which reaction (1) is at equilibrium, can be easily calculated using the van 't Hoff equation, yielding:

$$\left(\frac{P_{H2O}}{P_{H2}}\right)_{eq} = \exp\left(\frac{\Delta H_{TiO2} - \Delta H_{H2O}}{2RT}\right) \tag{2}$$

where $\Delta H_{TiO2} = -853$ kJ/mol O$_2$ is the enthalpy of TiO oxidation, i.e., of the reaction 2TiO + O$_2 \rightarrow$ 2TiO$_2$, and $\Delta H_{H2O} = -495$ kJ/mol O$_2$ is the enthalpy of water formation, 2H$_2$ + O$_2 \rightarrow$ 2H$_2$O. With these data and T = 673 K, Equation (2) yields $(P_{H2O}/P_{H2})_{eq} \approx 10^{-14}$. Since the water content due to impurities in the gas and to desorption from the reactor walls is certainly higher than that, we must expect that reaction (1) proceeds rightwards, i.e., that oxidation of TiO$_{1-\delta}$ NPs takes place during the thermal treatment. Notice that this behavior arises from the strongly negative formation enthalpy of TiO$_2$; oxides of late transition metals such as Fe and Co, on the contrary, would be reduced under the same atmosphere. This is a key factor in the processing of NCs samples, as we will show later on.

The oxidative power of the treatment atmosphere influences the rutile to anatase ratio. This is also in agreement with the literature, which shows that oxygen-deficient conditions favor the formation of rutile [30], because rutile itself exhibits a slightly oxygen-deficient stoichiometry.

3.2. Fe and Fe-Co Alloy NPs

Figure 5a shows the XRD patterns of the as-prepared Fe$_{100-x}$Co$_x$ NPs. The mean crystallite size d and lattice parameter of all identified phases are reported in Table 5. In all samples, we observe the (110) and (200) Bragg reflections characteristic of a body-centered cubic (BCC) α-phase. Only in the Co-richest sample Fe$_{32}$Co$_{68}$, the (111) and (200) Bragg reflections of a face-centered cubic (FCC) γ-phase are clearly visible. According to the Fe-Co phase diagram [31], Fe and Co are completely miscible at room temperature up to 72 at.% Co content forming a BCC α-phase, while FCC and BCC phases coexist at higher Co content (72 to 92 at.% Co). Here we observe the FCC γ-phase already at 68 at.% Co. This deviation with respect to the bulk phase diagram [31] in nanoalloy systems is due to the increase in relative magnitude of surface energy and is size-dependent [10]. FCC and BCC coexistence in Fe-Co nanoalloys has already been reported above 42 at.% Co for NPs of 15 nm [32].

Table 5. Crystallite size d and lattice parameter a of the three crystalline phases identified by XRD in Fe-Co NPs. The FCC γ-phase is only observed at the highest Co content.

Sample	α-Fe-Co		Cobalt Ferrite		γ-Fe-Co	
	d [nm]	a [Å]	d [nm]	a [Å]	d [nm]	a [Å]
Fe	15(1)	2.8729(4)	2.0(2)	8.451(6)	-	-
Fe$_{77}$Co$_{23}$	19(2)	2.8686(2)	2.7(2)	8.4272(6)	-	-
Fe$_{52}$Co$_{48}$	18(2)	2.8629(1)	2.4(2)	8.442(4)	-	-
Fe$_{32}$Co$_{68}$	24(2)	2.8406(2)	n.d.	n.d.	10(2)	3.564(2)

The (110) reflection of the α-phase (Figure 5b) shifts towards higher angles with increasing Co content. This corresponds to the shrinking of the lattice parameter a, as plotted in Figure 5c, and is consistent with the smaller atomic radius of Co with respect to Fe.

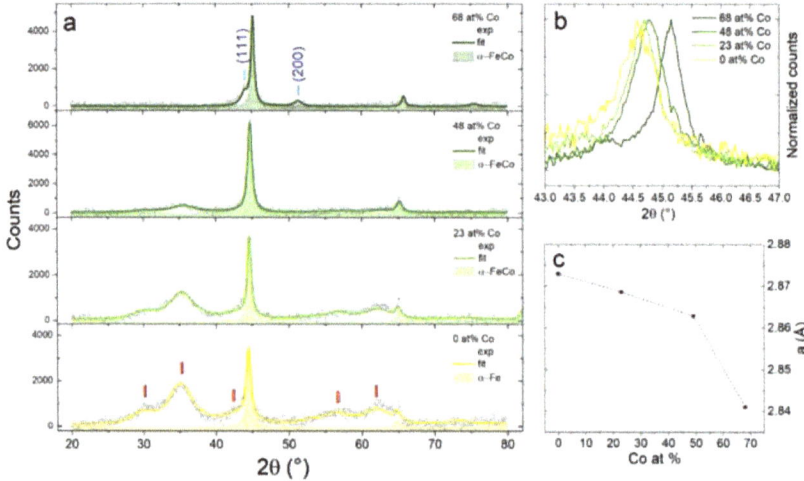

Figure 5. (a) XRD patterns of Fe$_{100-x}$Co$_x$ samples (x = 68–0 from top to bottom). Grey circles represent experimental data, while the colored solid lines are the best fits. The colored area highlights BCC peaks of the alloy, while colored vertical bars marks the Bragg reflections of cobalt ferrite (red) and FCC Fe-Co alloy (cyan); (b) A detail of the experimental data showing the shift of the (110) BCC reflection with increasing Co content; (c) Lattice parameter a obtained from Rietveld refinement (Table 5) as a function of Co content, with the error bars are displayed in red.

The broad peaks in the XRD patterns correspond to a magnetite (Fe$_3$O$_4$)-like spinel structure with an ultrafine crystallite size of 2–3 nm. This oxide forms when the NPs get in contact with the ambient air for XRD measurements. If air exposure is fast, the NPs begin to glow and oxidize completely. Conversely, if the exposure is sufficiently slow, as in this case where air flows into the sample holder gradually in about one hour, full oxidation can be avoided through the formation of a protective oxide shell [33,34]. The metal core / oxide shell morphology of the NPs gradually exposed to the air is confirmed by TEM investigations. The High Angle Annular Dark Field (HAADF) STEM image in Figure 6a shows a detail of Fe$_{52}$Co$_{48}$ sample. The NPs are surrounded by a lower contrast shell about 3 nm thick. This is a first indication in favor of the oxide nature of the shell. In fact, the contrast in incoherent HAADF-STEM images is proportional to $tZ^{1.7}$, where t is the thickness and Z is the average atomic number. Further information on the composition of the shell is provided by the STEM-EDX profile of a single NP recorded along the red line in Figure 6a. The shell corresponds to the regions, in which the Co and Fe fluorescence counts (blue and yellow areas) start to decrease while the O counts (red dashed line) are still constant. Interestingly, the Co/(Co + Fe) atomic ratio (black solid line) is similar in the core and the shell. This result demonstrates that the oxide is actually a cobalt ferrite (Fe$_{100-x}$Co$_x$)$_3$O$_4$ with a magnetite-like spinel structure [35] and a Co/Fe ratio similar to that of the core.

The HR-TEM image of the same NPs in Figure 6b highlights the atomic-level structure of core and shell regions. The Fast Fourier Transform (FFT) operated over the HR-TEM image is displayed in Figure 6c. The bright spots correspond to α-Fe-Co (110) planes, while the inner broad ring is due to the (311) crystalline planes of cobalt ferrite. These contributions can be separated by applying a filter to the FFT image followed by an inverse transformation. In this way, one obtains two separate images for the lattice planes of the two phases. In Figure 6d, the separate images are superposed in false colors to the HR-TEM image. The α-Fe-Co lattice planes (green) are clearly visible in the core, while the cobalt ferrite planes (violet) belong to the shell.

Finally, the intensity of the oxide broad peaks in Figure 5a shows that the oxidation resistance upon air exposure augments with increasing Co content. A remarkable stability against oxidation was also pointed out for Fe$_{100-x}$Co$_x$ NPs with $x \approx$ 40–50 prepared by hydrothermal synthesis [36].

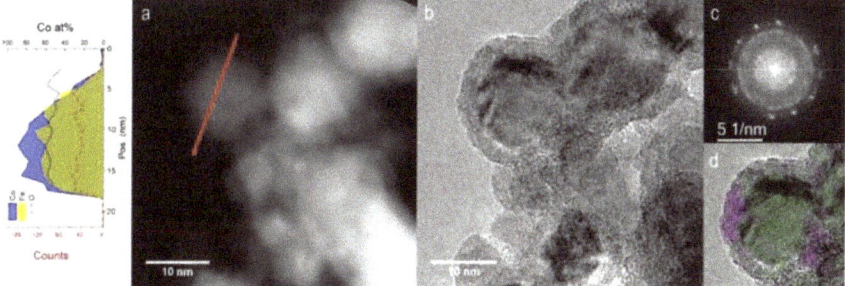

Figure 6. TEM analysis of Fe$_{52}$Co$_{48}$ NPs. (**a**) HAADF-STEM image and the corresponding EDX profile taken along the red arrow. The blue and yellow areas and the red dashed line represent Co, Fe and O X-ray fluorescence counts. The black solid line is the Co/(Co+Fe) atomic ratio within the NP; (**b**) HR-TEM image of the same area together with its (**c**) Fast-Fourier-Transform (FFT); (**d**) Mapping of crystal phases obtained from the analysis of planar spacing. α-FeCo (110) and cobalt ferrite (311) planes are represented in green and purple, respectively.

3.3. Fe/TiO$_x$ and Fe-Co/TiO$_x$ NCs

The STEM-EDX map in Figure 7 (superimposed to the corresponding STEM image) highlights the nanocomposite nature of the NPs assembly obtained by co-evaporation of Fe and Ti. The elemental distribution of Fe (yellow) and Ti (red) clearly shows the good intermixing at the nanoscale level of Fe-rich and Ti-rich NPs.

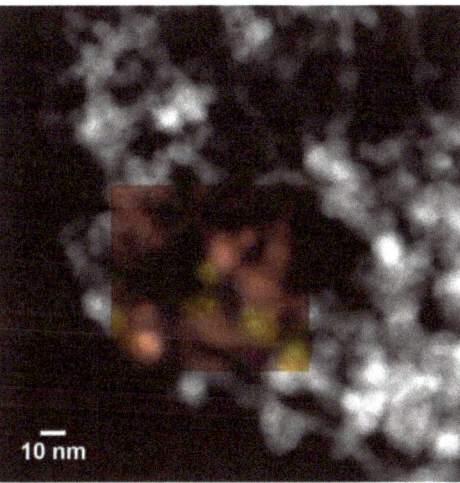

Figure 7. STEM-EDX elemental map of Fe (yellow) and Ti (red) in the as-prepared Fe/TiO$_x$ NC.

According to the XRD pattern (Figure 8a), the as-prepared sample is constituted by α-Fe and TiO$_{1-\delta}$. This view is confirmed by TEM analysis, which also provides further clues on the nanoscale phase distribution. The HR-TEM image in Figure 9 and the FFTs performed over the regions labeled a and b show the intimate contact between NPs constituted by α-Fe (region a) and TiO$_{1-\delta}$ (region b).

After a thermal treatment of 4 h at 400 °C under 1 MPa H$_2$ (Figure 8b), the intensity of TiO$_{1-\delta}$ Bragg reflections decreases significantly, while broad peaks attributable to rutile and anatase TiO$_2$ become visible. The crystallite size of α-Fe estimated from the breadth of XRD peaks is almost unaffected by the treatment, passing from 18 ± 2 nm of the as-prepared sample to 21 ± 2 nm. The STEM-EDX scan profile in Figure 10 shows an α-Fe NP of about 13 nm in contact with TiO$_x$ NPs.

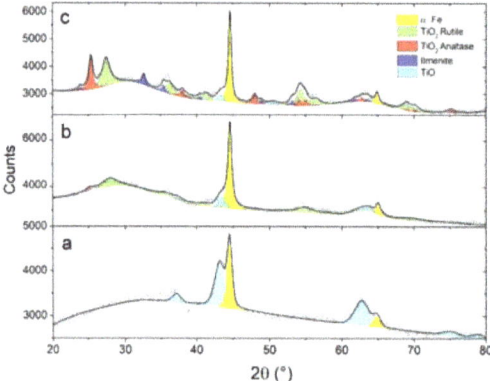

Figure 8. XRD patterns of Fe/TiO$_x$ NCs. (**a**) As-prepared sample; (**b,c**) After thermal treatment in a H$_2$ atmosphere, 1 MPa, for (**b**) 4 h and (**c**) 24 h. The colored areas highlight the contributions of different phases to the Rietveld refinement. Grey circles represent experimental data, while the black solid line shows the overall best fit.

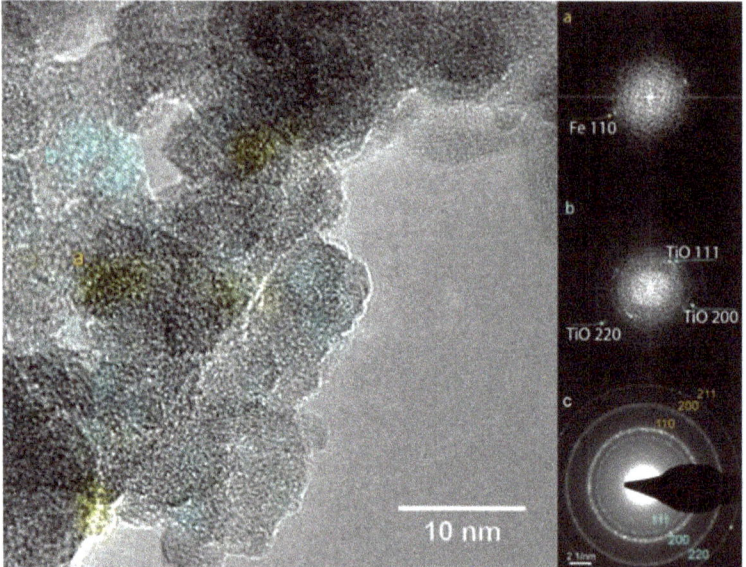

Figure 9. HR-TEM image of the as prepared Fe/TiO$_x$ NC (left) and FFTs performed on selected areas labeled with lowercase letters showing the presence of (**a**) Fe and (**b**) TiO$_{1-\delta}$ lattice planes; (**c**) SAD pattern: Fe and TiO$_{1-\delta}$ lattice planes are labeled in yellow and cyan, respectively (yellow and cyan rings serve as guides for the eye). In the HR-TEM image, the lattice planes of Fe and TiO$_{1-\delta}$ are highlighted using the same colors.

When the treatment is prolonged for 24 h, the XRD pattern clearly displays anatase and rutile TiO$_2$ Bragg reflections (Figure 8c) along with some residual TiO$_{1-\delta}$. In addition, the presence of a small amount of ilmenite FeTiO$_3$ is detected. The narrowing of the α-Fe peaks corresponds to an increased average crystallite size of about 30 nm. The HR-TEM image and phase mapping displayed in Figure 11 show an α-Fe NP in contact with both rutile and anatase, together with ilmenite in the interfacial region.

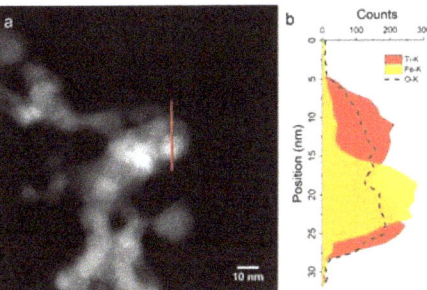

Figure 10. Fe/TiO$_x$ sample after 4 h in 1 MPa H$_2$ at 400 °C. (**a**) STEM image; (**b**) EDX line profile acquired along the red arrow, highlighting a Fe NP among TiO$_x$ NPs.

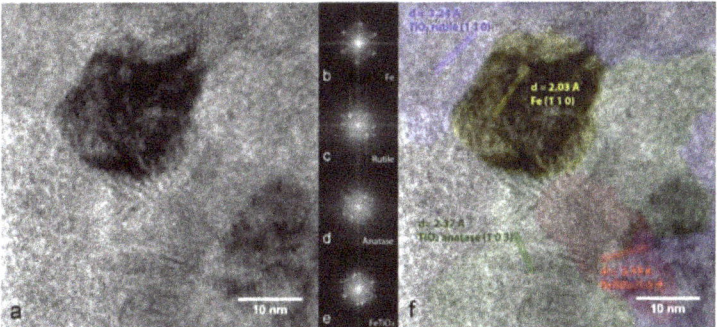

Figure 11. (**a**) HR-TEM image of the Fe/TiO$_x$ sample after 24 h in 1 MPa H$_2$ at 400 °C; (**b**,**e**) FFTs performed in selected areas as indicated in the crystalline phase distribution map in (**f**), which highlights the lattice planes of α-Fe (yellow) rutile (blue), anatase (green) and ilmenite (red).

The different oxidation thermodynamics of Fe and Ti explains the results of XRD. We already pointed out that oxidation of TiO$_{1-\delta}$ into TiO$_2$ is expected during the treatment due to water impurities. Considering now Fe, Equations (1) and (2) must be rewritten as:

$$(3/4)\text{Fe} + \text{H}_2\text{O} \leftrightarrow (1/4)\text{Fe}_3\text{O}_4 + \text{H}_2 \tag{3}$$

$$\left(\frac{P_{H2O}}{P_{H2}}\right)_{eq} = \exp\left(\frac{\Delta H_{Fe3O4} - \Delta H_{H2O}}{2RT}\right) \tag{4}$$

where $\Delta H_{Fe3O4} = -551$ kJ/mol O$_2$ is the enthalpy of the reaction $(3/2)$Fe + O$_2 \rightarrow (1/2)$Fe$_3$O$_4$. With these data and T = 673 K, equation (4) yields $(P_{H2O}/P_{H2})_{eq} \approx 7\cdot10^{-3}$. Since the water content in the H$_2$ atmosphere is likely below this value, reaction (3) proceeds leftwards, i.e., any Fe$_3$O$_4$ at the surface of the NPs should be reduced during the treatment. We notice that also the amount of Fe$_3$O$_4$ in the as-prepared sample is below the detection limit of XRD. It is possible that the contact with TiO$_{1-\delta}$ protects Fe NPs from oxidation during air exposure. In the case of Fe-Co alloy or pure Co NPs, the amount of oxide is expected to be even lower because the enthalpy of Co oxidation is slightly less negative compared to Fe, i.e., $\Delta H_{Co3O4} = -479$ kJ/mol O$_2$ and $\Delta H_{CoO} = -491$ kJ/mol O$_2$.

From the point of view of NPs mixing and TiO$_x$ stoichiometry, the scenario in Fe$_{50}$Co$_{50}$/TiO$_x$ is analogous to Fe/TiO$_x$. Figure 12 shows an EDX profile recorded for the as-prepared Fe$_{50}$Co$_{50}$/TiO$_x$ sample, in which a small Fe-Co NP with a diameter of about 7 nm appears supported on a Ti-rich NP. According to the previous discussion for Fe-Co NPs and Fe/TiO$_x$ NCs, we may expect that Fe$_{50}$Co$_{50}$ NPs crystallize in the BCC α-phase with some cobalt ferrite, while Ti-rich NPs develop mainly in the TiO$_{1-\delta}$ phase. The phase analysis by electron diffraction supports this view, as demonstrated in

Figure 13. In fact, the SAD azimuthal integration proves that the $Fe_{50}Co_{50}/TiO_x$ sample is composed by α-Fe-Co, cobalt ferrite and $TiO_{1-\delta}$. Cobalt ferrite is associated with the broad halos in the SAD pattern. This is compatible with the formation of a 2–3 nm thick oxide shell surrounding the metallic core, as shown in Section 3.2 for Fe-Co NPs.

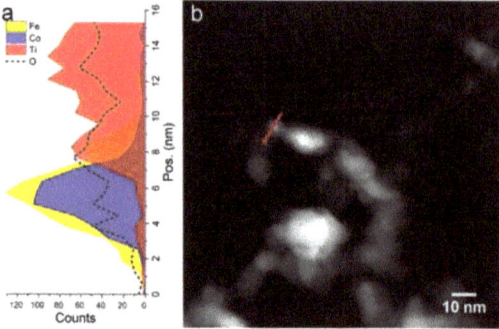

Figure 12. (a) EDX profile acquired along the path indicated with a red arrow in the corresponding STEM image (b) of the $Fe_{50}Co_{50}/TiO_x$ sample showing a Fe-Co NP in contact with TiO_x NPs.

A certain degree of NPs aggregation is typical of gas-phase condensation processes. If the NPs density in the gas phase is low (i.e., at low evaporation rates), aggregation takes place mainly on the collection surface. Welding of NPs is driven by capillary forces that act to reduce the surface free energy. At high evaporation rates, aggregation takes place also in the gas phase. The typical size of the aggregates is in the hundreds of nanometer range and their shape is generally much ramified. In elements with low melting points such as Mg, aggregation may even result in single crystal NPs [37]. In the presented metal/oxide nanocomposites, the oxide NPs seems to prevent aggregation of the metallic ones, as long as these are kept at a small volume fraction. This should allow to exploit interesting properties of individual NPs such as plasmonic resonance and superparamagnetism.

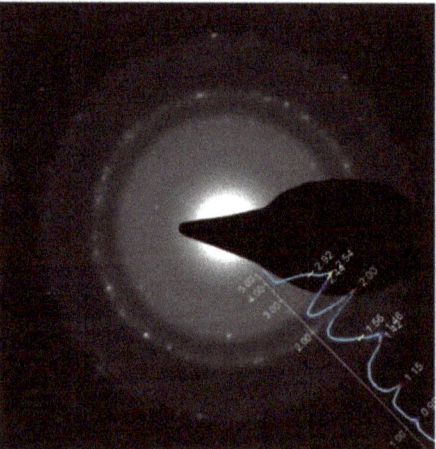

Figure 13. SAD acquired for the $Fe_{50}Co_{50}/TiO_x$ as-prepared sample. The overlaid blue plot is the profile analysis of the SAD obtained by azimuthal integration of the underlying image. The profile was computed by using the plugin PASAD [38] for the software Digital Micrograph from Gatan. The x-axis represents the interplanar spacing in Å. The position of Bragg reflections are indicated by the vertical colored bars and are labeled with the corresponding d-spacing. $TiO_{1-\delta}$, α-Fe-Co alloy, and cobalt ferrite (blue, red and yellow vertical bars respectively) are detected.

4. Conclusions

We have presented a novel synthesis method for the preparation of metal/oxide nanocomposites based on the physical assembly of nanoparticles (NPs), which are formed by gas phase condensation in a He/O_2 atmosphere. This approach goes beyond the simple post-synthesis partial oxidation of elemental [33,34] or alloy [39,40] NPs that can only yield a metal core-oxide shell morphology with obvious compositional restrictions. Indeed, our strategy has the potential to provide greater versatility in terms of both composition and independent control over the NPs size and morphology of the two phases.

The synthesis method was demonstrated here for Fe/TiO$_x$ and Fe-Co/TiO$_x$ nanocomposites, but can be extended to other metal/oxide combinations and may benefit from peculiar features of gas phase condensation, some of which were not explored in the present work. These include: (i) good nanoscale mixing can be achieved also in case of immiscible precursors [41]; (ii) the size of metal NPs can be controlled by tuning the evaporation rate and the inert gas pressure [37]; homogeneous alloy NPs can be synthesized provided that the evaporation rate of the elements are similar, for example Fe-Co, Fe-Ni, Co-Ni, Ag-Au, Au-Cu; NPs sources based on high-pressure sputtering can be employed for the synthesis of refractory metal and oxide NPs [39]; the NPs assembly can be compacted in situ to produce dense pellets with varying degrees of porosity [42]. Obtaining a metal/oxide nanocomposite from the evaporation of two metallic precursors requires that they exhibit strongly different oxidation enthalpies. Besides the case of Ti explored here, we envisage that other suitable precursors for the formation of oxide NPs may be Mg, Al, and Si, all having an oxidation enthalpy more negative than -900 kJ/mol O_2. These may be combined with NPs of late transition metals, including noble metals. Post-synthesis thermal treatments in a suitable atmosphere permit to control the stoichiometry of the oxide NPs to a certain extent, reducing at the same time the oxidized surface shell around metal NPs. Future work will explore other metal/oxide combinations and characterize their physical/chemical properties.

Author Contributions: Conceptualization and methodology, N.P. and L.P.; investigation, N.P. and A.M.; project administration and resources, L.P. and V.M.; writing—original draft preparation, N.P. and L.P.; writing—review & editing, all authors; supervision, L.P.

Funding: This research received no external funding.

Conflicts of Interest: The authors declare no conflicts of interest.

References

1. Hernández Mejía, C.; van Deelen, T.W.; de Jong, K.P. Activity enhancement of cobalt catalysts by tuning metal-support interactions. *Nat. Commun.* **2018**, *9*, 4459. [CrossRef] [PubMed]
2. Kattel, S.; Liu, P.; Chen, J.G. Tuning Selectivity of CO_2 Hydrogenation Reactions at the Metal/Oxide Interface. *J. Am. Chem. Soc.* **2017**, *139*, 9739–9754. [CrossRef] [PubMed]
3. Mutschler, R.; Moioli, E.; Luo, W.; Gallandat, N.; Züttel, A. CO_2 hydrogenation reaction over pristine Fe, Co, Ni, Cu and Al_2O_3 supported Ru: Comparison and determination of the activation energies. *J. Catal.* **2018**, *366*, 139–149. [CrossRef]
4. Suchorski, Y.; Kozlov, S.M.; Bespalov, I.; Datler, M.; Vogel, D.; Budinska, Z.; Neyman, K.M.; Rupprechter, G. The role of metal/oxide interfaces for long-range metal particle activation during CO oxidation. *Nat. Mater.* **2018**, *17*, 519–522. [CrossRef] [PubMed]
5. Zheng, N.; Stucky, G.D. A General Synthetic Strategy for Oxide-Supported Metal Nanoparticle Catalysts. *J. Am. Chem. Soc* **2006**, *128*, 14278–14280. [CrossRef] [PubMed]
6. Prieto, G.; Zečević, J.; Friedrich, H.; De Jong, K.P.; De Jongh, P.E. Towards stable catalysts by controlling collective properties of supported metal nanoparticles. *Nat. Mater.* **2013**, *12*, 34–39. [CrossRef] [PubMed]
7. White, R.J.; Luque, R.; Budarin, V.L.; Clark, J.H.; Macquarrie, D.J. Supported metal nanoparticles on porous materials. Methods and applications. *Chem. Soc. Rev.* **2009**, *38*, 481–494. [CrossRef]
8. Zhang, Z.; Zhang, L.; Hedhili, M.N.; Zhang, H.; Wang, P. Plasmonic Gold Nanocrystals Coupled with Photonic Crystal Seamlessly on TiO_2 Nanotube Photoelectrodes for Efficient Visible Light Photoelectrochemical Water Splitting. *Nano Lett.* **2013**, *13*, 14–20. [CrossRef]

9. Lu, A.-H.; Salabas, E.L.; Schüth, F. Magnetic Nanoparticles: Synthesis, Protection, Functionalization, and Application. *Angew. Chemie Int. Ed.* **2007**, *46*, 1222–1244. [CrossRef]
10. Calvo, F. Thermodynamics of nanoalloys. *Phys. Chem. Chem. Phys.* **2015**, *17*, 27922–27939. [CrossRef]
11. Henry, C.R. Morphology of supported nanoparticles. *Prog. Surf. Sci.* **2005**, *80*, 925–116. [CrossRef]
12. Farmer, J.A.; Campbell, C.T. Ceria maintains smaller metal catalyst particles by strong metal-support bonding. *Science* **2010**, *329*, 933–936. [CrossRef] [PubMed]
13. Vayssilov, G.N.; Lykhach, Y.; Migani, A.; Staudt, T.; Petrova, G.P.; Tsud, N.; Skála, T.; Bruix, A.; Illas, F.; Prince, K.C.; et al. Support nanostructure boosts oxygen transfer to catalytically active platinum nanoparticles. *Nat. Mater.* **2011**, *10*, 310–315. [CrossRef]
14. Nilsson, A.; Pettersson, L.; Nørskov, J.K. *Chemical Bonding at Surfaces and Interfaces*; Elsevier: New York, NY, USA, 2008; ISBN 0080551912.
15. Xu, B.-Q.; Wei, J.-M.; Yu, Y.-T.; Li, Y.; Li, J.-L.; Zhu, Q.-M. Size Limit of Support Particles in an Oxide-Supported Metal Catalyst: Nanocomposite Ni/ZrO$_2$ for Utilization of Natural Gas. *J. Phys. Chem. B* **2003**, *107*, 5203–5207. [CrossRef]
16. Joo, S.H.; Park, J.Y.; Tsung, C.-K.; Yamada, Y.; Yang, P.; Somorjai, G.A. Thermally stable Pt/mesoporous silica core-shell nanocatalysts for high-temperature reactions. *Nat. Mater.* **2009**, *8*, 126–131. [CrossRef] [PubMed]
17. Farrusseng, D.; Tuel, A. Perspectives on zeolite-encapsulated metal nanoparticles and their applications in catalysis. *New J. Chem.* **2016**, *40*, 3933–3949. [CrossRef]
18. Ennas, G.; Marongiu, G.; Marras, S.; Piccaluga, G. Mechanochemical Route for the Synthesis of Cobalt Ferrite-Silica and Iron-Cobalt Alloy-Silica Nanocomposites. *J. Nanoparticle Res.* **2004**, *6*, 99–105. [CrossRef]
19. Ennas, G.; Falqui, A.; Marras, S.; Sangregorio, C.; Marongiu, G. Influence of Metal Content on Size, Dispersion, and Magnetic Properties of Iron-Cobalt Alloy Nanoparticles Embedded in Silica Matrix. *Chem. Mater.* **2004**, *16*, 5659–5663. [CrossRef]
20. Huang, Y.L.; Xue, D.S.; Zhou, P.H.; Ma, Y.; Li, F.S. α-Fe-Al$_2$O$_3$ nanocomposites prepared by sol-gel method. *Mater. Sci. Eng. A* **2003**, *359*, 332–337. [CrossRef]
21. Freund, H.-J. Clusters and islands on oxides: from catalysis via electronics and magnetism to optics. *Surf. Sci.* **2002**, *500*, 271–299. [CrossRef]
22. Ischia, G.; Wenk, H.-R.; Lutterotti, L.; Berberich, F. Quantitative Rietveld texture analysis of zirconium from single synchrotron diffraction images. *J. Appl. Crystallogr.* **2005**, *38*, 377–380. [CrossRef]
23. Nesmeyanov, A.N. *Vapor Pressure of the Chemical Elements*; Academic Press: New York, NY, USA, 1963.
24. Patelli, N.; Calizzi, M.; Migliori, A.; Morandi, V.; Pasquini, L. Hydrogen Desorption Below 150 °C in MgH$_2$-TiH$_2$ Composite Nanoparticles: Equilibrium and Kinetic Properties. *J. Phys. Chem. C* **2017**, *121*, 11166–11177. [CrossRef]
25. Gusev, A.I.; Valeeva, A.A. The influence of imperfection of the crystal lattice on the electrokinetic and magnetic properties of disordered titanium monoxide. *Phys. Solid State* **2003**, *45*, 1242–1250. [CrossRef]
26. Xu, J.; Wang, D.; Yao, H.; Bu, K.; Pan, J.; He, J.; Xu, F.; Hong, Z.; Chen, X.; Huang, F. Nano Titanium Monoxide Crystals and Unusual Superconductivity at 11 K. *Adv. Mater.* **2018**, *30*, 1706240. [CrossRef] [PubMed]
27. Semaltianos, N.G.; Logothetidis, S.; Frangis, N.; Tsiaoussis, I.; Perrie, W.; Dearden, G.; Watkins, K.G. Laser ablation in water: A route to synthesize nanoparticles of titanium monoxide. *Chem. Phys. Lett.* **2010**, *496*, 113–116. [CrossRef]
28. Rossi, G.; Calizzi, M.; Di Cintio, V.; Magkos, S.; Amidani, L.; Pasquini, L.; Boscherini, F. Local Structure of V Dopants in TiO$_2$ Nanoparticles: X-ray Absorption Spectroscopy, Including Ab-Initio and Full Potential Simulations. *J. Phys. Chem. C* **2016**, *120*, 7457–7466. [CrossRef]
29. Binetti, E.; Koura, Z.E.; Patel, N.; Dashora, A.; Miotello, A. Rapid hydrogenation of amorphous TiO$_2$ to produce efficient H-doped anatase for photocatalytic water splitting. *Appl. Catal. A-Gen.* **2015**, *500*, 69–73. [CrossRef]
30. Hanaor, D.A.H.; Sorrell, C.C. Review of the anatase to rutile phase transformation. *J. Mater. Sci.* **2011**, *46*, 855–874. [CrossRef]
31. Johnson, E.C.; Ridout, M.S.; Cranshaw, T.E. The Mossbauer Effect in Iron Alloys. *Proc. Phys. Soc.* **1963**, *81*, 6. [CrossRef]
32. Wang, Z.H.; Choi, C.J.; Kim, J.C.; Kim, B.K.; Zhang, Z.D. Characterization of Fe-Co alloyed nanoparticles synthesized by chemical vapor condensation. *Mater. Lett.* **2003**, *57*, 3560–3564. [CrossRef]

33. Signorini, L.; Pasquini, L.; Savini, L.; Carboni, R.; Boscherini, F.; Bonetti, E.; Giglia, A.; Pedio, M.; Mahne, N.; Nannarone, S. Size-dependent oxidation in iron/iron oxide core-shell nanoparticles. *Phys. Rev. B* **2003**, *68*, 195423. [CrossRef]
34. Pasquini, L.; Barla, A.; Chumakov, A.I.; Leupold, O.; Rüffer, R.; Deriu, A.; Bonetti, E. Size and oxidation effects on the vibrational properties of nanocrystalline α-Fe. *Phys. Rev. B* **2002**, *66*, 073410. [CrossRef]
35. Nlebedim, I.C.; Moses, A.J.; Jiles, D.C. Non-stoichiometric cobalt ferrite, CoxFe$_3$ − xO$_4$ (x = 1.0 to 2.0): Structural, magnetic and magnetoelastic properties. *J. Magn. Magn. Mater.* **2013**, *343*, 49–54. [CrossRef]
36. Klencsár, Z.; Németh, P.; Sándor, Z.; Horváth, T.; Sajó, I.E.; Mészáros, S.; Mantilla, J.; Coaquira, J.A.H.; Garg, V.K.; Kuzmann, E.; et al. Structure and magnetism of Fe-Co alloy nanoparticles. *J. Alloys Compd.* **2016**, *674*, 153–161. [CrossRef]
37. Venturi, F.; Calizzi, M.; Bals, S.; Perkisas, T.; Pasquini, L. Self-assembly of gas-phase synthesized magnesium nanoparticles on room temperature substrates. *Mater. Res. Express* **2015**, *2*, 1. [CrossRef]
38. Gammer, C.; Mangler, C.; Rentenberger, C.; Karnthaler, H.P. Quantitative local profile analysis of nanomaterials by electron diffraction. *Scr. Mater.* **2010**, *63*, 312–315. [CrossRef]
39. Grammatikopoulos, P.; Steinhauer, S.; Vernieres, J.; Singh, V.; Sowwan, M. Nanoparticle design by gas-phase synthesis. *Adv. Phys. X* **2016**, *6149*, 1–20. [CrossRef]
40. Blackmore, C.E.; Rees, N.V.; Palmer, R.E. Modular construction of size-selected multiple-core Pt-TiO$_2$ nanoclusters for electro-catalysis. *Phys. Chem. Chem. Phys.* **2015**, *17*, 28005–28009. [CrossRef]
41. Calizzi, M.; Venturi, F.; Ponthieu, M.; Cuevas, F.; Morandi, V.; Perkisas, T.; Bals, S.; Pasquini, L. Gas-phase synthesis of Mg-Ti nanoparticles for solid-state hydrogen storage. *Phys. Chem. Chem. Phys.* **2016**, *18*, 141–148. [CrossRef]
42. Tanimoto, H.; Pasquini, L.; Prümmer, R.; Kronmüller, H.; Schaefer, H.-E. Self-diffusion and magnetic properties in explosion densified nanocrystalline Fe. *Scr. Mater.* **2000**, *42*. [CrossRef]

© 2019 by the authors. Licensee MDPI, Basel, Switzerland. This article is an open access article distributed under the terms and conditions of the Creative Commons Attribution (CC BY) license (http://creativecommons.org/licenses/by/4.0/).

Article

Effective La-Na Co-Doped TiO$_2$ Nano-Particles for Dye Adsorption: Synthesis, Characterization and Study on Adsorption Kinetics

Inderjeet Singh and Balaji Birajdar *

Special Centre for Nano Sciences, Jawaharlal Nehru University, New Delhi 110067, India; lathaphysics69@gmail.com
* Correspondence: birajdar@mail.jnu.ac.in; Tel.: +91-11-26704743

Received: 20 October 2018; Accepted: 13 December 2018; Published: 9 March 2019

Abstract: The mesoporous La-Na co-doped TiO$_2$ nanoparticles (NPs) have been synthesized by non-aqueous, solvent-controlled, sol-gel route. The substitutional doping of large sized Na^{+1} and La^{+3} at Ti^{4+} is confirmed by X-ray diffraction (XRD) and further supported by Transmission Electron Microscopy (TEM) and X-ray Photo-electron Spectroscopy (XPS). The consequent increase in adsorbed hydroxyl groups at surface of La-Na co-doped TiO$_2$ results in decrease in pH$_{IEP}$, which makes nanoparticle surface more prone to cationic methylene blue (MB) dye adsorption. The MB dye removal was examined by different metal doping, pH, contact time, NPs dose, initial dye concentration and temperature. Maximum dye removal percentage was achieved at pH 7.0. The kinetic analysis suggests adsorption dynamics is best described by pseudo second-order kinetic model. Langmuir adsorption isotherm studies revealed endothermic monolayer adsorption of Methylene Blue dye.

Keywords: La-Na co-doped TiO$_2$; non-aqueous solvent controlled sol-gel route; physical adsorption; methylene; blue

---

1. Introduction

A leading source of water pollution is dye containing effluents from industries like textile [1], printing [2], leather [3], pharmaceuticals [4] and kraft bleaching [5] and so forth. Many of these dyes are mutagenic, carcinogenic and even lead to chromosomal fractures [6–8] causing health hazards to living beings. Eliminating such dyes from industrial effluents is becoming increasingly necessary. Prominent methods used for this are: adsorption [9–11], coagulation/flocculation [12,13], membrane filtration [14,15] and so forth. Among these, adsorption methods are advantageous because they are simple, economical and effective. In addition, adsorption efficiency can be controlled by many factors including adsorbent surface area, adsorbent dose, pH, contact time and adsorbate concentration [16–18].

Recently nanostructured material such as TiO$_2$ [19,20], magnetic iron oxides [21–23] and nanoparticle loaded carbon [24,25] have been reported by several researchers for efficient removal of organic dyes. Due to their large surface area and easy tailoring of surface properties, use of nanomaterials can be very effective in adsorbing dyes from industrial effluents. The pH$_{IEP}$ (the pH at which zeta potential of nanoparticle equals zero) plays a vital role in the adsorption kinetics of dye at nanoparticle surface because it affects formation of bonds between surface hydroxyl groups and dye molecules during chemisorption. Hence, modification of nanoparticle surface can be a good strategy for physical removal of organic dye. TiO$_2$ is well studied material, which allows easy tailoring of its surface by doping and varying synthesis methods. Alkali dopants [26–31] have been long used to enhance surface area of TiO$_2$ nanoparticles and creation of new surface sites which, significantly improve the adsorption of dyes. On the other hand, it is reported that minute doping of rare earth elements in TiO$_2$ [32–35] not only concentrate the organic dye at nanoparticle surface but also stabilizes the meso-structure of TiO$_2$.

Therefore modification of TiO$_2$ using co-doping of rare earth elements and alkali metals is expected to be more effective in the adsorption of dye at the surface. In view of this, pure, La doped, Na doped and La-Na co-doped TiO$_2$ nanopowder were prepared by solvent controlled non aqueous sol-gel route [36–38]. A systematic investigation was carried out to understand the role of dopants on adsorption property of dye at nanoparticle surface. Earlier reports [26,39] demonstrated that Na$^+$ (1.02 Å) and La^{+3} (1.03 Å), due to their large size, could not enter in TiO$_2$ lattice to substitute for Ti^{4+} (0.68 Å) but migrates to TiO$_2$ surface. In present work, substitutional doping of Na and La is confirmed by XRD, TEM and XPS for the first time. The variation in pH$_{IEP}$ values with doping is well explained and correlated with adsorption behavior of dye at modified TiO$_2$ surface.

2. Experimental Section

2.1. Synthesis Method

Non-aqueous, solvent-controlled, sol-gel route is excellent for preparation of highly pure oxide nanoparticles (NPs) with good yield and is therefore adopted for the synthesis of TiO$_2$ NPs. Pure TiO$_2$ gel was prepared by mixing 20 mL Titanium tetra isopropoxide (Spectrochem, Mumbai, India) with 40 mL methyl cellosolve (SRL Chem, Mumbai, India) by stirring for 2 h at pH 3 (maintained using 1 M HNO$_3$ (Merck, Darmstadt, Germany). Under similar conditions, gels of La doped TiO$_2$, Na doped TiO$_2$ and La-Na co-doped TiO$_2$ were prepared using lanthanum nitrate (SRL, Mumbai, India) and sodium nitrate (CDH, Mumbai, India) as La and Na precursor respectively. The labels and nominal composition of samples are indexed in Table 1. The prepared gels were dried under Infrared IR lamp and grinded to fine powder. For crystallization, all powders were calcined at 450 °C for 1 h. The substantial concentrations of metal doping were determined by wavelength dispersive X-ray fluorescence spectroscopy (WDXRF) and obtained results are also included in Table 1.

Table 1. Nominal & actual dopant concentration, 2θ values and crystallite size of NPs.

NPs Labelling	Nominal Dopant Conc. (at. %)		Dopant Conc. (at. %)		2θ (Degree) A(101)	Crystallite Size (nm)
	Na	La	Na	La		
PT	0	0	0	0	25.722	14.0
LT	0	1	0	0.90	25.512	8.0
NT4	4	0	3.89	0	25.466	10.0
LNT2	2	1	1.82	0.81	25.425	7.5
LNT4	4	1	3.75	0.76	25.359	7.0
LNT6	6	1	5.67	0.74	25.380	9.0

2.2. Characterization

The concentration of metal dopant was identified by wavelength dispersive X-ray fluorescence spectroscopy (WDXRF) (Bruker S4 PIONEER, Texas, USA). X-ray diffraction (XRD) patterns were acquired on Rigaku diffractometer (Cu K$_\alpha$, λ = 0.154 nm) to determine crystallite size. The Brunauer-Emmett-Teller (BET) analysis of NP's was performed using Quantachrome Instrument, Florida, USA after degassing at 300 °C for 4 h. JEOL 2100-F (Tokyo, Japan) TEM operating at 200 kV was used for transmission electron microscopy (TEM) analysis. Transmission electron microscope (TEM) images, high resolution TEM (HRTEM) images, selective area electron diffraction (SAED) pattern, Scanning TEM (STEM) images & energy dispersive X-ray (EDX) spectra were acquired to determine crystallite size, d-spacing, crystallinity and elemental analysis. The binding energy of each element present in sample was determined via X-ray photoelectron spectroscopy (XPS) (XPS oxford instrument using X-rays of energy 1486.6 eV, Concord, MA, USA). Zeta potential analyzer (ZEECOM Microtec, Tokyo, Japan) was used to determine isoelectric point (IEP) of the prepared nanopowder.

2.3. Adsorption Studies

Methylene blue (Hi media, Pune, India) dye adsorption at NPs surface was carried out by batch technique. All the adsorption experiments were carried out at room temperature (25 °C). The effect of varying metal doping, initial pH, contact time, adsorbent NPs dose, initial dye concentration and temperature on adsorption was studied. In general, a particular amount of NPs dose is added in 10 mL methylene blue (MB) dye solution and stirred for required time. The pH of the solution is adjusted by using 0.01 M NaOH and 0.01 M HCl. The adsorption capacity q_e (mg/g) and removal percentage of prepared NPs for MB dye were calculated by the Equations (1) and (2) [25]:

$$q_e = (C_0 - C_e)V/m \tag{1}$$

$$\text{Removal \%} = ((C_0 - C_t)/C_0) \times 100 \tag{2}$$

where, C_0 (mg/L), C_e (mg/L) and C_t (mg/L) are the MB dye concentrations at initial time, equilibrium time and contact time 't' respectively. V (L) is the total volume of the suspension and m (g) is mass of adsorbent NPs. In order to measure concentration of dye at different contact time, NPs are removed from the solution by centrifugation and absorbance of supernatant measured by UV-Visible spectrometer. These absorbance results are calibrated with standard samples and then concentration of residual dye is measured. For data consistency, all the adsorption experiments are performed in duplicates.

3. Results and Discussion

3.1. XRD Analysis

Structural characterization of nanopowder was carried out by XRD and is reported in Figure 1. XRD patterns of PT show the diffraction peaks of pure TiO_2 anatase phase. Similar peaks are observed for LT, NT4, LNT2, LNT4 and LNT6. Absence of extra peaks confirms the formation of pure anatase phase in doped NPs and that there are no segregated phases of dopants. The shifts in peak position with doping confirm the substitutional doping of large sized La^{+3} (1.03 Å) and Na^{+1} (1.02 Å) dopant at Ti^{4+} (0.68 Å) site. Generally, doping of large ionic radii metal ions in host lattice TiO_2 results in strain in crystal structure. This lattice strain is compensated by formation of oxygen vacancies [38,40]. Secondly the substitution of host lattice ion by low valence metal ion is also likely to yield oxygen vacancies [41]. Thus, doping of large sized and low valent Na or La metal ions in TiO_2 matrix would induce formation of oxygen vacancies. Besides, La and Na mono-doping and co-doping in TiO_2 reduces the intensity of X-ray peaks of anatase phase. As compared to Na doping, La doping shows stronger reduction in the intensity of anatase phase. This could be attributed to fewer oxygen vacancies expected upon La doping as compared to Na due to high valence state of La.

The crystal plane (101) was selected to determine average crystallite size of prepared nano-powders using Debye Scherrer formula. The calculated average crystallite size and 2θ values are tabulated in Table 1. Clearly, the average crystallite size decreases for individual as well as co-doped TiO_2 nano-powder. Besides, there is critical doping concentration (LNT4) after which crystallite size increases. The decrease in crystallite size is obviously due to substitution of low valent and large sized dopants resulting in strained crystal structure and hence oxygen vacancies [41,42].

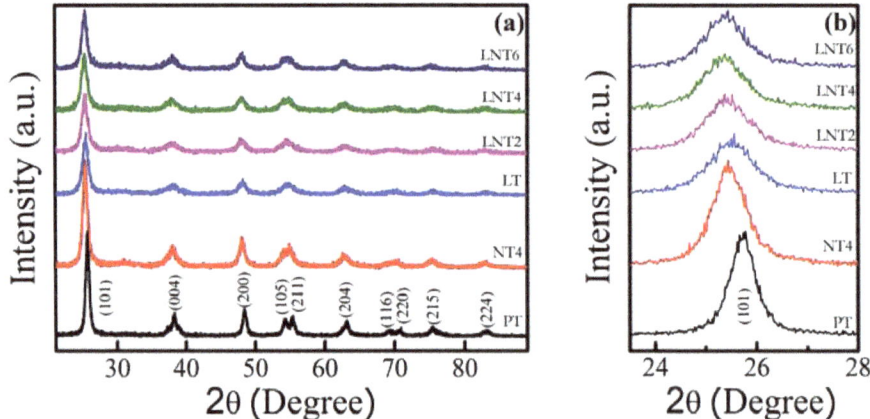

Figure 1. (a) XRD patterns of NPs. (b) Amplified image of (a).

3.2. BET Surface Areas and Pore Size

The mesoporous structure and surface area of prepared samples were examined by isothermal curves formed by adsorption and desorption of N_2 and the Barrett-Joyner-Halenda (BJH) pore size distribution curves. As shown in Figure 2a, all prepared NPs exhibited H2 type hysteresis loop with adsorption-desorption isotherms of type 4, typical characteristic of mesoporous structure with ink-bottle shaped pores [43,44]. The respective pore size distribution curves (Figure 2b) are obtained from desorption isotherm of pure and doped TiO_2 samples using BJH method. The BET surface area, pore volume and pore diameter of all the samples are tabulated in Table 2.

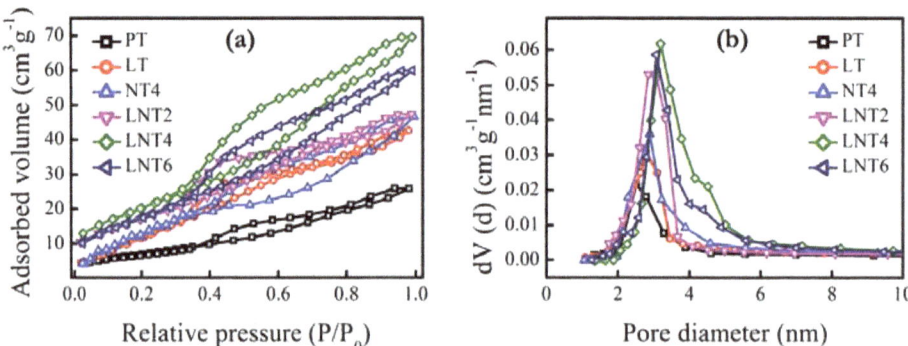

Figure 2. (a) N_2 adsorption-desorption isotherms and (b) pore diameter distribution curve of prepared NPs.

Table 2. The pH_{IEP}, BET surface area, BJH pore volume and pore diameter of all synthesized NPs.

NPs	pH_{IEP}	BET Surface Area ($m^2\ g^{-1}$)	Pore Volume ($cm^3\ g^{-1}$)	Pore Diameter (nm)
PT	5.7	45.712	0.039	2.32
LT	5.5	54.342	0.052	2.59
NT4	5.2	49.629	0.066	2.70
LNT2	5.0	63.418	0.080	2.85
LNT4	4.7	74.996	0.119	3.19
LNT6	4.6	67.971	0.104	3.08

Clearly, there is variation in hysteresis loop and pore distribution curve with doping. Increase in La and Na doping in pure TiO_2 results in large surface area due to decreased crystallite size. In addition, pore size and hence pore volume increases with certain level of dopant concentration as in LNT4 and thereafter decreases. This could be attributed to formation of oxygen vacancies [45]. Moreover, this speculation is also supported by high porosity of NT4 sample as compared to LT, because NT4 possess more oxygen vacancies.

3.3. Morphology

TEM, HRTEM and SAED give structural and morphological information about prepared nanopowder. TEM bright field images of Figure 3a PT and Figure 3b LNT4 further corroborate XRD result that crystallite size of LNT4 is smaller than PT nanopowder. The representative particle size of LNT 4 nanopowder (8–12 nm) is smaller than PT nanopowder (15–20 nm). In addition, SAED pattern (Inset of Figure 3a,b) form ring patterns which are indexed according to anatase phase of TiO_2 which confirm the crystalline phase of PT and LNT4. The substitutional doping of dopants is further confirmed by increased d spacing in LNT4 (Figure 3d) as compared to that of PT (Figure 3c).

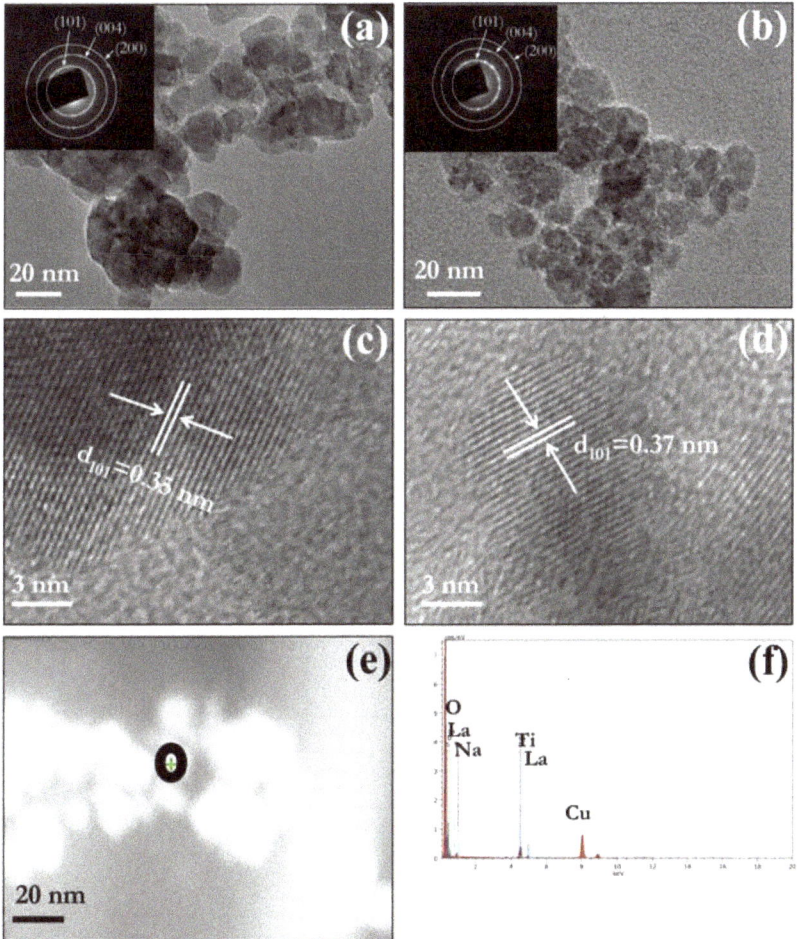

Figure 3. TEM micrograph of (**a**) PT and (**b**) LNT4 nanopowder. Insets in (**a**,**b**) show SAED patterns of PT and LNT4 respectively. HRTEM images of (**c**) PT and (**d**) LNT4. (**e**) STEM dark field image of cluster of LNT4 nanopowder. (**f**) STEM-EDX point spectrum from an area shown by black circle in (**e**).

Figure 3e shows STEM dark field image of cluster of LNT4 nanopowder. The simultaneous presence of La, Na and Ti peak in point EDX spectrum (Figure 3f) taken from the area shown by a black circle (Figure 3e) corroborate XRD and HRTEM results.

3.4. XPS Analysis

The binding energy of different elements present in prepared samples, were examined by XPS analysis. The obtained spectra were baseline corrected and then fitted by commonly used Voigt function. As shown in Figure 4, the high resolution O 1s XPS spectra of all samples composed of two peaks corresponding to Ti-O link and adsorbed hydroxyl groups (Ti–OH link) [46]. The increase in peak intensity of Ti-OH link with Na and La doping confirms increased adsorption of hydroxyl groups at the surface. This is due to increased formation of oxygen vacancies [47] at the surface of nanoparticle due to substitutional doping of La and Na at Ti site (confirmed by XRD). Furthermore, substitutional doping of La and Na at Ti site is also confirmed by comparative study of high resolution XPS spectra of Ti 2p (Figure S1, Supplementary File). The doublet of Ti 2p of PT is composed of Ti $2p_{1/2}$ (B.E. 464.4 eV) and Ti $2p_{3/2}$ (B.E. 458.7 eV) and indicates Ti^{+4} state. The binding energy shows red shift (Table 3) with La and Na doping, which could be attributed to the lower electronegativity of La (1.10) and Na (0.93) than that of Ti (1.52). This result confirms substitutional doping of La and Na at Ti site [48] and absence of other oxidation state of Ti.

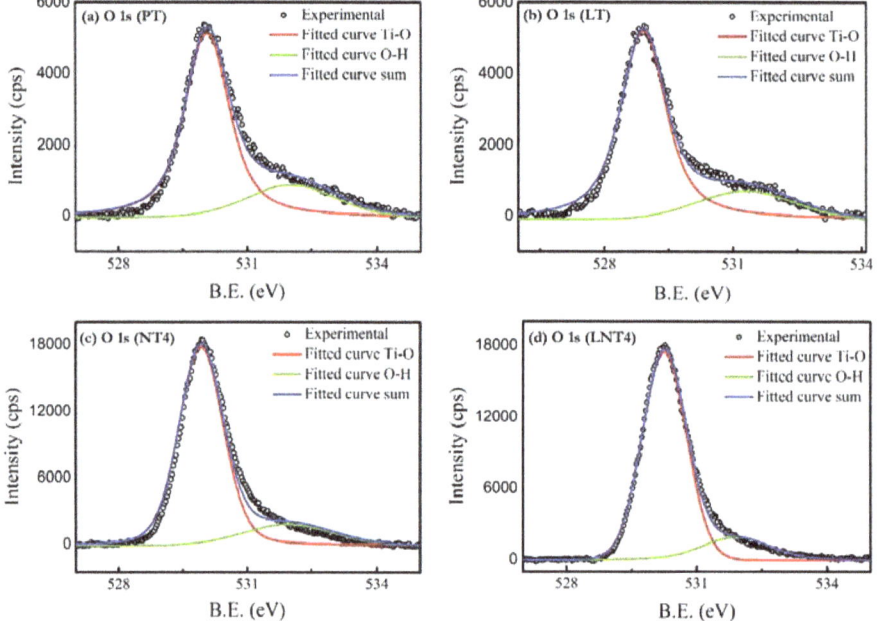

Figure 4. The High resolution XPS spectra of O 1s for (**a**) PT, (**b**) LT, (**c**) NT4 and (**d**) LNT4.

Table 3. Peak position of different peaks obtained from high resolution XPS of different samples.

Sample	Ti-O Peak	O-H Peak	Ti $2p_{3/2}$ Peak	Ti $2p_{1/2}$ Peak	La $3d_{5/2}$ Peak 1	La $3d_{5/2}$ Peak 2	La $3d_{3/2}$ Peak 1	La $3d_{3/2}$ Peak 2	Na 1s Peak
PT	530.07	532.03	458.71	464.41	-	-	-	-	-
LT	529.93	531.96	458.46	464.16	834.99	839.18	851.51	856.04	-
NT4	528.91	531.29	458.19	464.06	-	-	-	-	1071.34
LNT4	529.90	531.91	457.75	463.38	834.83	839.03	851.38	855.89	1071.23

In addition, the high resolution XPS spectra of La 3d (Figure S2, Supplementary File) and Na 1s (Figure S3, Supplementary File) confirm the simultaneous presence of La and Na. The La 3d peak composed of multiplet $3d_{5/2}$ and $3d_{3/2}$ with core level binding energy 834.9 eV and 851.8 eV respectively corresponding to La^{+3} oxidation state [49]. The distance between two peaks $La3d_{5/2}$ (peak 1) and La $3d_{3/2}$ (peak 1) corresponds to bonding of La with O [50]. Similarly, Na 1s peak with core level binding energy 1071.3 eV represents Na^{+1} oxidation state [51].

3.5. Zeta Potential Study and Isoelectric Point

The surface charge of nanoparticles play a vital role on adsorption kinetics of dye at surface of nanoparticles. Zeta potential measurement technique was used to determine surface charge. The pH value at which there is no charge at the surface and hence zero zeta potential is defined as isoelectric point. The surface of suspended TiO_2 nanoparticles in water covered with hydroxyl groups [52] and therefore surface charge is function of pH of solution according to Equations (3) and (4),

$$Ti^{+4}-OH + H^+ \rightarrow Ti^{+4}-OH_2^+ \quad (3)$$

$$Ti^{+4}-OH \rightarrow Ti^{+4}-O^- + H^+ \quad (4)$$

When pH of solution is less than pH_{IEP}, nanoparticle surface is positively charged according to Equation (3) and favors adsorption of anionic species. Whereas, when pH of solution is greater than pH_{IEP}, nanoparticle surface is negatively charged according to Equation (4) resulting in adsorption of cationic species.

Figure 5A shows the zeta potential values of different nanopowder as function of pH of solution. It is clear from Figure 5B that pH_{IEP} value decreases from PT to LNT6. This is attributed to increased surface area and adsorbed hydroxyl groups [45,52]. The increased surface area results in more adsorption of hydroxyl groups and hence more hydrogen ions are produced, which decreases pH_{IEP} value [53]. A higher concentration of hydroxyl groups at the surface of nanoparticle is expected to favor more adsorption of dye molecules.

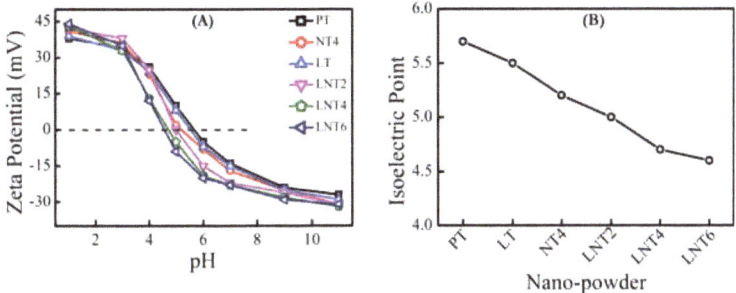

Figure 5. (A) Effect of doping in TiO_2 on dispersion zeta potential at different potential. (B) The dispersion isoelectronic point (IEP) for different NPs.

3.6. Adsorption Studies

Effect of metal doping. The effect La and Na doping in TiO$_2$ on adsorption of MB dye was studied at optimal conditions [equilibrium time = 15 min; pH = 7.0; dye conc. = 5 mg/L; NPs dose = 1.2 g/L] by considering absorbance peak at 664 nm. At pH 7, all NPs have negative zeta potential (Figure 5A) which facilitates adsorption of cationic MB dye. It is clear from Figure 6a that LNT4 NPs shows best removal rate of MB dye, even if LNT2, LNT4 and LNT6 possess nearly equal negative zeta potential. This is attributed to small crystallite size and large surface area of LNT4 as compared to LNT2 and LNT6.

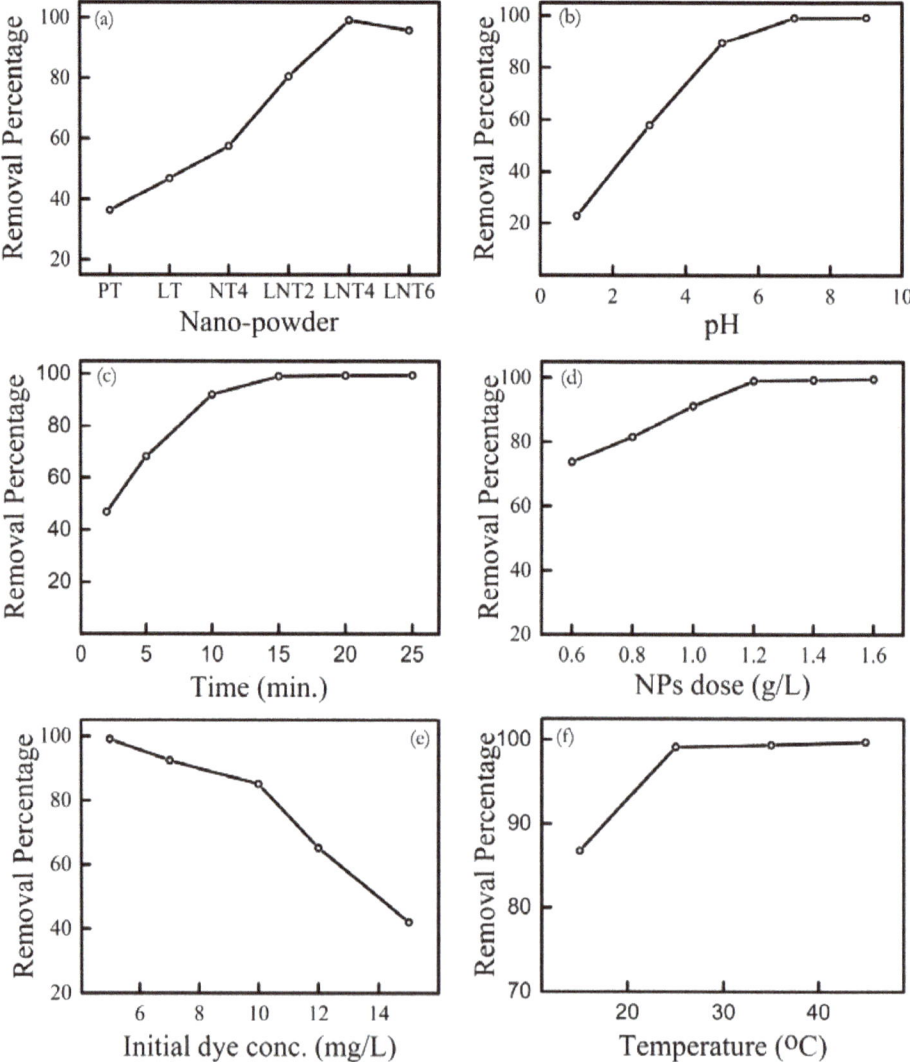

Figure 6. (a) Effect of metal doping in TiO$_2$ on dye removal. (b–e) Dye removal by adsorption for LNT4: (b) Effect of solution pH. (c) Effect of contact time. (d) Effect of LNT4 NPs dose. (e) Effect of initial dye concentration. (f) Effect of temperature.

Effect of initial pH. The LNT4 NPs shows best dye removal percentage among all prepared NPs. Therefore, further adsorption experiments are performed using LNT4 NPs as adsorbate. The absorbance spectra of MB dye at different pH is shown in Figure S4 (supplementary file). There is no change in peak position and shape of absorbance spectra observed. Figure 6b shows effect of pH on dye removal percentage by LNT4 NPs [equilibrium time = 15 min; dye conc. =5 mg/L; NPs dose = 1.2 g/L]. At lower pH, LNT4 shows very less dye removal percentage, which is obvious due to positive zeta potential resulting in repulsive force on cationic dye. However, when pH is greater than pH$_{IEP}$, dye removal percentage increases. For example, at pH 7.0, 99% of the dye is removed. This is because of electrical attraction between cationic MB dye and negative surface charge of LNT4 at pH > pH$_{IEP}$. At pH more than 7.0 there is negligible increase in removal percentage. Therefore, all further sorption experiments were conducted at pH 7.0.

Effect of contact time. The effect of contact time on dye adsorption at LNT4 NPs surface was investigated at different time intervals (2, 5, 10, 15, 20 and 25 min) at optimal conditions [pH = 7.0; dye conc. =5 mg/L; NPs dose = 1.2 g/L]. Dye removal percentage increased from 47% to 99% with increase in contact time from 2 min to 25 min. The sorption equilibrium was achieved within 15 min. Attaining equilibrium in such short span of time indicates chemical adsorption of dye, instead of physical adsorption that takes relatively long contact time [54]. After 15 min, there is negligible uptake of dye is observed (Figure 6c). Therefore, further adsorption experiments were carried out for 15 min.

Effect of LNT4 NPs dose. The effect of LNT4 NPs dose on MB dye uptake is shown in Figure 6d. MB dye sorption experiments were conducted using 5 mg/L dye solution at pH 7.0 [equilibrium time = 15 min]. The LNT4 NPs dose varied from 0.6 g/L to 1.6 g/L. MB dye removal percentage increased considerably from 73% to 99% on increasing dose from 0.6 g/L to 1.2 g/L. This increase in removal percentage with increase in adsorbent dose is obvious and attributed more available surface area and active surface sites [55]. No further uptake of MB dye was observed on increasing NPs dose to 1.4 g/L and 1.6 g/L. Therefore, 1.2 g/L dose was chosen as optimum dose for further adsorption experiments.

Effect of initial dye concentration. The effect of initial dye concentration on dye sorption was investigated using 5–15 mg/L MB dye concentration [equilibrium time =15 min; pH =7.0; NPs dose =1.2 g/L]. The dye removal percentage drops from 99% to 42% as dye concentration is increased from 5 mg/L to 15 mg/L (Figure 6e) due to less available surface area and surface sites. Similar results were reported by many authors [16,56,57].

Effect of temperature. Generally wastewater effluents from industries are hot and hence adsorbents used must be stable and thermal resistant [58]. Therefore, investigation of dye adsorption at different temperature is important. Figure 6f shows the removal rate of MB dye at different temperature by LNT4 NPs [equilibrium time =15 min; pH =7.0; NPs dose =1.2 g/L; dye conc.= 5 mg/L]. Clearly there is more adsorption of MB dye with rise in temperature, which implies endothermic adsorption process. The favorable increase in adsorption of dye with rise in temperature can be explained by two ways [10]: (i) elevated interaction between cationic MB dye and hydroxyl groups at surface of NPs; (ii) enhanced diffusion of MB dye molecules within pores of NPs. In addition, there may be disruption of agglomeration with rise in temperature, which facilitates increase in surface area and hence increased dye adsorption. Many reports [57,59] investigated effect of temperature on adsorption of dye at adsorbent surface and similar results were found.

3.7. Adsorption Kinetics

In order to investigate the mechanism of adsorption, several kinetic models are used to fit experimental data. Most commonly used kinetic models, namely pseudo first order, pseudo second order and intra-particle diffusion are used to determine preferential kinetic model for dye adsorption. The best fit model is determined by comparing linear regression correlation coefficient (R^2) value. The simple form of pseudo-first-order, pseudo-second-order and intra-particle diffusion model might be represented by Equations (5)–(7) respectively.

$$\log(q_e - q_t) = \log(q_e) - (k_1/2.303)t \tag{5}$$

$$(t/q_t) = 1/k_2 q_e^2 + t/q_e \tag{6}$$

$$q_t = k_i t^{1/2} + C \tag{7}$$

where k_1 (min^{-1}), k_2 (g mg^{-1} min^{-1}) and k_i (mg g^{-1} min$^{-1/2}$) are rate constants of pseudo-first-order, pseudo-second-order reaction and intra-particle diffusion respectively and C is the intercept. q_e and q_t are the adsorption capacity at equilibrium and at contact time 't' respectively. The rate constant k_1 could be determined from slope of linear plot log $(q_e - q_t)$ versus t. Similarly, slope of linear plot of t/q_t versus t gives the rate constant k_2. The rate constant k_i can be evaluated from the slope of the plot of q_t versus $t^{1/2}$.

The plot of pseudo first order, second order and intra-particle diffusion kinetic model obtained for adsorption of MB dye at the surface of LNT4 NPs is shown in Figure 7A–C. The obtained regression coefficient (R^2) and calculated k values are listed in Table 4. The value of regression coefficient (R^2) of second order kinetic model is higher than first order, which could be attributed to heterogeneous nature of TiO$_2$ NPs [23]. Thus the kinetics of MB dye adsorption on LNT4 NPs is best described by pseudo second order kinetic model.

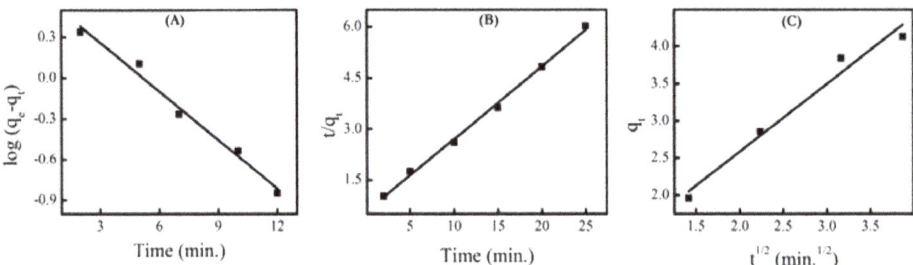

Figure 7. (**A**) pseudo first order, (**B**) pseudo second order and (**C**) intra-particle diffusion kinetic plots for adsorption of dye at LNT4 NPs surface.

Table 4. Adsorption kinetics parameter obtained from adsorption of MB dye at surface of LNT4 NPs.

1st Order Kinetics		2nd Order Kinetics		Intra-Particle Diffusion	
k_1	R^2	k_2	R^2	k_i	R^2
0.253	0.9796	0.081	0.9959	0.762	0.9616

The slope of intra-particle diffusion plot measure rate of intra-particle diffusion, which can occur after external surface adsorption. Whereas, intercept of plot is proportional to boundary layer thickness and measure the contribution of surface adsorption to rate controlling step [60]. The low value of regression coefficient (Figure 7c) reflects less involvement of intra-particle diffusion to rate controlling step.

3.8. Adsorption Isotherm

The binding mechanism of dye at NPs surface can be understood by analyzing adsorption isotherm. The adsorption isotherm of MB dye at surface of NPs was studied using Langmuir model, which consider adsorbent surface to be mono-layered with uniform energy at which adsorption occurs [21]. The Langmuir isotherm equation is given by Equation (8) [61] as:

$$C_e/q_e = 1/bq_{max} + C_e/q_{max} \qquad (8)$$

where, b (L/mg) is Langmuir isotherm adsorption constant and q_{max} (mg/g) is adsorption capacity, which are calculated using slope and intercept, obtained from linear plot of C_e/q_e versus C_e (Figure 8).

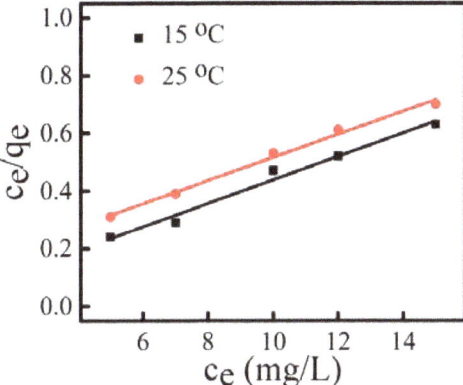

Figure 8. Langmuir isotherm plot for adsorption of MB dye at two different temperatures.

The Langmuir adsorption constants and correlation coefficient (R^2) for adsorption of MB dye at two different temperatures are given in Table 5. The value of R^2 is higher than 0.9, which confirms that monolayer adsorption of dye is predominant. In addition, the value of q_{max} calculated by Langmuir isotherm model increases with rise in temperature indicating adsorption process is endothermic.

Table 5. Adsorption isotherm parameter obtained from adsorption of MB dye at surface of LNT4 NPs.

Langmuir Isotherm Constant	15 °C	25 °C
q_{max}	24.62	25.04
b	1.35	0.33
R^2	0.97	0.99

4. Conclusions

Physical removal of organic cationic MB dye by La-Na co doped TiO_2 nano-powder prepared by non-aqueous, solvent-controlled, sol-gel route is demonstrated. The prepared NPs are crystalline and mesoporous as confirmed by XRD and BET analysis. Furthermore, XRD and TEM results confirm the substitution of large sized Na^{+1} and La^{+3} at Ti^{+4} sites. This low valent metal ion doping results in formation of oxygen vacancies which facilitates more adsorption of hydroxyl groups at the surface of NPs (confirmed by XPS). The adsorbed hydroxyl groups reduce the pH_{IEP} value and, therefore facilitate effective adsorption of cationic MB dye. The adsorption capacity of LNT4 NPs is found to be highest, which could be attributed to its high surface area and porosity. In addition, adsorption kinetics of MB dye at surface of LNT4 NPs is best described by pseudo second-order kinetic model due to heterogeneous nature of titania NPs. Langmuir adsorption isotherm studies revealed endothermic monolayer adsorption of Methylene Blue dye.

Supplementary Materials: The following are available online at http://www.mdpi.com/2079-4991/9/3/400/s1, Figure S1: High resolution XPS spectra of Ti 2p of prepared samples, Figure S2: High resolution XPS spectra of La 3d of prepared samples, Figure S3: High resolution XPS spectra of Na 1s of prepared samples, Figure S4:

(a) Effect of pH on absorbance spectra of MB dye. (b) Amplified image of (a) at pH 7 and pH9, Table S1: Area under different peaks obtained from high resolution XPS of different samples, Table S2: FWHM of different peaks obtained from high resolution XPS of different samples.

Author Contributions: Idea, Synthesis and experiments by I.S. Data analysis and manuscript preparation by I.S. and B.B.

Funding: The authors are grateful for the financial support by the UPE-II (project ID 102) and DST purse-II grant.

Acknowledgments: Financial assistance by Jawaharlal Nehru University (JNU) for the publication of this article is gratefully acknowledged. We kindly acknowledge SPS, JNU, New Delhi and AIRF, JNU, New Delhi for access to XRD and TEM respectively. We also acknowledge BIT, Bangalore and MNIT, Jaipur for access to BET analysis and XPS respectively. The authors are thankful to Himadri Bohidar, School pf Physical Science, JNU New Delhi for discussion on agglomeration of mesoporous NPs. I.S. acknowledges UGC, India for financial support via senior research fellowship (SRF).

Conflicts of Interest: The authors declare no conflict of interest.

References

1. Robinson, T.; McMullan, G.; Marchant, R.; Nigam, P. Remediation of dyes in textile effluent: A critical review on current treatment technologies with a proposed alternative. *Bioresour. Technol.* **2001**, *77*, 247–255. [CrossRef]
2. Vachon, S.; Klassen, R.D. Green project partnership in the supply chain: The case of the package printing industry. *J. Clean. Prod.* **2006**, *14*, 661–671. [CrossRef]
3. Tünay, O.; Kabdasli, I.; Orhon, D.; Ates, E. Characterization and pollution profile of leather tanning industry in Turkey. *Water Sci. Technol.* **1995**, *32*, 1–9. [CrossRef]
4. Kolpin, D.W.; Furlong, E.T.; Meyer, M.T.; Thurman, E.M.; Zaugg, S.D.; Barber, L.B.; Buxton, H.T. Pharmaceuticals, hormones, and other organic wastewater contaminants in U.S. streams, 1999–2000: A national reconnaissance. *Environ. Sci. Technol.* **2002**, *36*, 1202–1211. [CrossRef] [PubMed]
5. Zhang, Q.; Chuang, K.T. Adsorption of organic pollutants from effluents of a Kraft pulp mill on activated carbon and polymer resin. *Adv. Environ. Res.* **2001**, *5*, 251–258. [CrossRef]
6. Yagub, M.T.; Sen, T.K.; Afroze, S.; Ang, H.M. Dye and its removal from aqueous solution by adsorption: A review. *Adv. Colloid Interface Sci.* **2014**, *209*, 172–184. [CrossRef] [PubMed]
7. Hildenbrand, S.; Schmahl, F.W.; Wodarz, R.; Kimmel, R.; Dartsch, P.C. Azo dyes and carcinogenic aromatic amines in cell cultures. *Int. Arch. Occup. Environ. Health* **1999**, *72*, M052–M056. [CrossRef]
8. Gupta, V.K.; Nayak, A.; Bhushan, B.; Agarwal, S. A critical analysis on the efficiency of activated carbons from low-cost precursors for heavy metals remediation. *Crit. Rev. Environ. Sci. Technol.* **2015**, *45*, 613–668. [CrossRef]
9. Ramakrishna, K.; Viraraghavan, T. Dye removal using low cost adsorbents. *Water Sci. Technol.* **1997**, *36*, 189–196. [CrossRef]
10. Doğan, M.; Alkan, M.; Türkyilmaz, A.; Özdemir, Y. Kinetics and mechanism of removal of methylene blue by adsorption onto perlite. *J. Hazard. Mater.* **2004**, *109*, 141–148. [CrossRef]
11. Gupta, V.K.; Kumar, R.; Nayak, A.; Saleh, T.A.; Barakat, M.A. Adsorptive removal of dyes from aqueous solution onto carbon nanotubes: A review. *Adv. Colloid Interface Sci.* **2013**, *193–194*, 24–34. [CrossRef] [PubMed]
12. Sadri Moghaddam, S.; Alavi Moghaddam, M.R.; Arami, M. Coagulation/flocculation process for dye removal using sludge from water treatment plant: Optimization through response surface methodology. *J. Hazard. Mater.* **2010**, *175*, 651–657. [CrossRef] [PubMed]
13. Verma, A.K.; Dash, R.R.; Bhunia, P. A review on chemical coagulation/flocculation technologies for removal of colour from textile wastewaters. *J. Environ. Manag.* **2012**, *93*, 154–168. [CrossRef] [PubMed]
14. Van Der Bruggen, B.; Lejon, L.; Vandecasteele, C. Reuse, treatment, and discharge of the concentrate of pressure-driven membrane processes. *Environ. Sci. Technol.* **2003**, *37*, 3733–3738. [CrossRef]
15. Stoquart, C.; Servais, P.; Bérubé, P.R.; Barbeau, B. Hybrid Membrane Processes using activated carbon treatment for drinking water: A review. *J. Memb. Sci.* **2012**, *411–412*, 1–12. [CrossRef]
16. Gupta, V.K.; Nayak, A. Cadmium removal and recovery from aqueous solutions by novel adsorbents prepared from orange peel and Fe_2O_3 nanoparticles. *Chem. Eng. J.* **2012**, *180*, 81–90. [CrossRef]

17. Nayak, A.; Bhushan, B.; Gupta, V.; Sharma, P. Chemically activated carbon from lignocellulosic wastes for heavy metal wastewater remediation: Effect of activation conditions. *J. Colloid Interface Sci.* **2017**, *493*, 228–240. [CrossRef] [PubMed]
18. Gupta, V.K.; Ali, I.; Saleh, T.A.; Nayak, A.; Agarwal, S. Chemical treatment technologies for waste-water recycling—An overview. *RSC Adv.* **2012**, *2*, 6380–6388. [CrossRef]
19. Mahmoodi, N.M.; Hayati, B.; Arami, M.; Bahrami, H. Preparation, characterization and dye adsorption properties of biocompatible composite (alginate/titania nanoparticle). *Desalination* **2011**, *275*, 93–101. [CrossRef]
20. Abou-Gamra, Z.M.; Ahmed, M.A. TiO_2 Nanoparticles for Removal of Malachite Green Dye from Waste Water. *Adv. Chem. Eng. Sci.* **2015**, *5*, 373–388. [CrossRef]
21. Zeng, S.; Duan, S.; Tang, R.; Li, L.; Liu, C.; Sun, D. Magnetically separable $Ni_{0.6}Fe_{2.4}O_4$ nanoparticles as an effective adsorbent for dye removal: Synthesis and study on the kinetic and thermodynamic behaviors for dye adsorption. *Chem. Eng. J.* **2014**, *258*, 218–228. [CrossRef]
22. Iram, M.; Guo, C.; Guan, Y.; Ishfaq, A.; Liu, H. Adsorption and magnetic removal of neutral red dye from aqueous solution using Fe_3O_4 hollow nanospheres. *J. Hazard. Mater.* **2010**, *181*, 1039–1050. [CrossRef] [PubMed]
23. Saha, B.; Das, S.; Saikia, J.; Das, G. Preferential and enhanced adsorption of different dyes on iron oxide nanoparticles: A comparative study. *J. Phys. Chem. C* **2011**, *115*, 8024–8033. [CrossRef]
24. Narayanasamy, L.; Murugesan, T. Degradation of Alizarin Yellow R using UV/H_2O_2 Advanced Oxidation Process. *Environ. Sci. Technol.* **2014**, *33*, 482–489. [CrossRef]
25. Asfaram, A.; Ghaedi, M.; Hajati, S.; Goudarzi, A. Ternary dye adsorption onto MnO_2 nanoparticle-loaded activated carbon: Derivative spectrophotometry and modeling. *RSC Adv.* **2015**, *5*, 72300–72320. [CrossRef]
26. Xie, H.; Li, N.; Liu, B.; Yang, J.; Zhao, X. Role of sodium ion on TiO_2 photocatalyst: Influencing crystallographic properties or serving as the recombination center of charge carriers? *J. Phys. Chem. C* **2016**, *120*, 10390–10399. [CrossRef]
27. Bessekhouad, Y.; Robert, D.; Weber, J.V.; Chaoui, N. Effect of alkaline-doped TiO_2 on photocatalytic efficiency. *J. Photochem. Photobiol. A Chem.* **2004**, *167*, 49–57. [CrossRef]
28. Yang, G.; Yan, Z.; Xiao, T.; Yang, B. Low-temperature synthesis of alkalis doped TiO_2 photocatalysts and their photocatalytic performance for degradation of methyl orange. *J. Alloys Compd.* **2013**, *580*, 15–22. [CrossRef]
29. Panagiotopoulou, P.; Kondarides, D.I. Effects of alkali promotion of TiO_2 on the chemisorptive properties and water-gas shift activity of supported noble metal catalysts. *J. Catal.* **2009**, *267*, 57–66. [CrossRef]
30. Chen, L.C.; Huang, C.M.; Tsai, F.R. Characterization and photocatalytic activity of K^+-doped TiO_2 photocatalysts. *J. Mol. Catal. A Chem.* **2007**, *265*, 133–140. [CrossRef]
31. López, T.; Hernandez-Ventura, J.; Gómez, R.; Tzompantzi, F.; Sánchez, E.; Bokhimi, X.; García, A. Photodecomposition of 2,4-dinitroaniline on Li/TiO_2 and Rb/TiO_2 nanocrystallite sol-gel derived catalysts. *J. Mol. Catal. A Chem.* **2001**, *167*, 101–107. [CrossRef]
32. Liqiang, J.; Xiaojun, S.; Baifu, X.; Baiqi, W.; Weimin, C. The preparation and characterization of La doped TiO_2 nanoparticles and their photocatalytic activity. *J. Solid State Chem.* **2004**, *177*, 3375–3382. [CrossRef]
33. Dai, K.; Peng, T.; Chen, H.; Liu, J.; Zan, L. Photocatalytic Degradation of Commercial Phoxim over La-Doped TiO_2 Nanoparticles in Aqueous Suspension. *Environ. Sci. Technol.* **2009**, *43*, 1540–1545. [CrossRef] [PubMed]
34. Uzunova-Bujnova, M.; Todorovska, R.; Dimitrov, D.; Todorovsky, D. Lanthanide-doped titanium dioxide layers as photocatalysts. *Appl. Surf. Sci.* **2008**, *254*, 7296–7302. [CrossRef]
35. Xu, A.-W.; Gao, Y.; Liu, H.-Q. The Preparation, Characterization, and their Photocatalytic Activities of Rare-Earth-Doped TiO_2 Nanoparticles. *J. Catal.* **2002**, *207*, 151–157. [CrossRef]
36. Garnweitner, G.; Niederberger, M. Nonaqueous and surfactant-free synthesis routes to metal oxide nanoparticles. *J. Am. Ceram. Soc.* **2006**, *89*, 1801–1808. [CrossRef]
37. Niederberger, M.; Garnweitner, G.; Ba, J.; Polleux, J.; Pinna, N. Nonaqueous synthesis, assembly and formation mechanisms of metal oxide nanocrystals. *Int. J. Nanotechnol.* **2007**, *4*, 263–281. [CrossRef]
38. Singh, I.; Kumar, R.; Birajdar, B.I. Zirconium doped TiO_2 nano-powder via halide free non-aqueous solvent controlled sol-gel route. *J. Environ. Chem. Eng.* **2017**, *5*, 2955–2963. [CrossRef]
39. Liu, J.W.; Han, R.; Wang, H.T.; Zhao, Y.; Lu, W.J.; Wu, H.Y.; Yu, T.F.; Zhang, Y.X. Degradation of PCP-Na with La-B co-doped TiO_2 series synthesized by the sol-gel hydrothermal method under visible and solar light irradiation. *J. Mol. Catal. A Chem.* **2011**, *344*, 145–152. [CrossRef]

40. Yu, J.C.; Lin, J.; Kwok, R.W.M. $Ti_{1-x}Zr_xO_2$ Solid Solutions for the Photocatalytic Degradation of Acetone in Air. *J. Phys. Chem. B* **1998**, *102*, 5094–5098. [CrossRef]
41. Pan, X.; Yang, M.-Q.; Fu, X.; Zhang, N.; Xu, Y.-J. Defective TiO_2 with oxygen vacancies: Synthesis, properties and photocatalytic applications. *Nanoscale* **2013**, *5*, 3601. [CrossRef] [PubMed]
42. Deshpande, S.; Patil, S.; Kuchibhatla, S.V.; Seal, S. Size dependency variation in lattice parameter and valency states in nanocrystalline cerium oxide. *Appl. Phys. Lett.* **2005**, *87*, 1–3. [CrossRef]
43. Kleitz, F.; Bérubé, F.; Guillet-Nicolas, R.; Yang, C.M.; Thommes, M. Probing adsorption, pore condensation, and hysteresis behavior of pure fluids in three-dimensional cubic mesoporous KIT-6 silica. *J. Phys. Chem. C* **2010**, *114*, 9344–9355. [CrossRef]
44. Kleitz, F.; Liu, D.; Anilkumar, G.M.; Park, I.-S.; Solovyov, L.A.; Shmakov, A.N.; Ryoo, R. Large Cage Face-Centered-Cubic *Fm3m* Mesoporous Silica: Synthesis and Structure. *J. Phys. Chem. B* **2003**, *107*, 14296–14300. [CrossRef]
45. Singh, I.; Birajdar, B. Synthesis, characterization and photocatalytic activity of mesoporous Na-doped TiO_2 nano-powder prepared via a solvent-controlled non-aqueous sol-gel route. *RSC Adv.* **2017**, *7*, 54053–54062. [CrossRef]
46. Simmons, G.W.; Beard, B.C. Characterization of Acid-Base Properties of the Hydrated Oxides on Iron and Titanium Metal Surfaces. *J. Phys. Chem.* **1987**, *91*, 1143–1148. [CrossRef]
47. Jing, L.; Xin, B.; Yuan, F.; Xue, L.; Wang, B.; Fu, H. Effects of surface oxygen vacancies on photophysical and photochemical processes of Zn-doped TiO_2 nanoparticles and their relationships. *J. Phys. Chem. B* **2006**, *110*, 17860–17865. [CrossRef]
48. Wang, J.; Yu, Y.; Li, S.; Guo, L.; Wang, E.; Cao, Y. Doping behavior of Zr^{4+} ions in Zr^{4+}-doped TiO_2 nanoparticles. *J. Phys. Chem. C* **2013**, *117*, 27120–27126. [CrossRef]
49. Yu, Y.-G.; Chen, G.; Hao, L.-X.; Zhou, Y.-S.; Pei, J.; Sun, J.-X.; Han, Z.-H. Doping La into the depletion layer of the $Cd_{0.6}Zn_{0.4}S$ photocatalyst for efficient H_2 evolution. *Chem. Commun.* **2013**, *49*, 10142–10144. [CrossRef]
50. Sunding, M.F.; Hadidi, K.; Diplas, S.; Løvvik, O.M.; Norby, T.E.; Gunnæs, A.E. XPS characterisation of in situ treated lanthanum oxide and hydroxide using tailored charge referencing and peak fitting procedures. *J. Electron Spectrosc. Relat. Phenom.* **2011**, *184*, 399–409. [CrossRef]
51. Guo, Z.; Zhou, J.; An, L.; Jiang, J.; Zhu, G.; Deng, C. A new-type of semiconductor $Na_{0.9}Mg_{0.45}Ti_{3.55}O_8$: preparation, characterization and photocatalysis. *J. Mater. Chem. A Mater. Energy Sustain.* **2014**, *2*, 20358–20366. [CrossRef]
52. Suttiponparnit, K.; Jiang, J.; Sahu, M.; Suvachittanont, S. Role of Surface Area, Primary Particle Size, and Crystal Phase on Titanium Dioxide Nanoparticle Dispersion Properties | Nanoscale Research Letters | Full Text. *Nanoscale Res. Lett.* **2011**, *6*, 1–8.
53. Yu, J.C.; Lin, J.; Kwok, R.W.M. Enhanced photocatalytic activity of $Ti_{1-x}V_xO_2$ solid solution on the degradation of acetone. *J. Photochem. Photobiol. A Chem.* **1997**, *111*, 199–203. [CrossRef]
54. Malik, P.K. Use of activated carbons prepared from sawdust and rice-husk for adsoprtion of acid dyes: A case study of acid yellow 36. *Dye. Pigment.* **2003**, *56*, 239–249. [CrossRef]
55. Garg, V.K.; Gupta, R.; Kumar, R.; Gupta, R.K. Adsorption of chromium from aqueous solution on treated sawdust. *Bioresour. Technol.* **2004**, *89*, 121–124. [CrossRef]
56. Moussavi, G.; Mahmoudi, M. Removal of azo and anthraquinone reactive dyes from industrial wastewaters using MgO nanoparticles. *J. Hazard. Mater.* **2009**, *168*, 806–812. [CrossRef] [PubMed]
57. Dogan, M.; Alkan, M.; Onganer, Y. Adsorption of Methylene Blue from Aqueous Solution onto Perlite. *Water Air Soil Pollut.* **2000**, *120*, 229–248. [CrossRef]
58. Homaeigohar, S.; Zillohu, A.U.; Abdelaziz, R.; Hedayati, M.K.; Elbahri, M. A novel nanohybrid nanofibrous adsorbent for water purification from dye pollutants. *Materials* **2016**, *9*, 848. [CrossRef]
59. Asfour, H.M.; Fadali, O.A.; Nassar, M.M.; El-Geundi, M.S. Equilibrium studies on adsorption of basic dyes on hardwood. *J. Chem. Technol. Biotechnol.* **1985**, *35A*, 21–27. [CrossRef]

60. Bahgat, M.; Farghali, A.A.; El Rouby, W.; Khedr, M.; Mohassab-Ahmed, M.Y. Adsorption of methyl green dye onto multi-walled carbon nanotubes decorated with Ni nanoferrite. *Appl. Nanosci.* **2013**, *3*, 251–261. [CrossRef]
61. Bagbi, Y.; Sarswat, A.; Mohan, D.; Pandey, A.; Solanki, P.R. Lead and Chromium Adsorption from Water using L-Cysteine Functionalized Magnetite (Fe$_3$O$_4$) Nanoparticles. *Sci. Rep.* **2017**, 1–15. [CrossRef]

© 2019 by the authors. Licensee MDPI, Basel, Switzerland. This article is an open access article distributed under the terms and conditions of the Creative Commons Attribution (CC BY) license (http://creativecommons.org/licenses/by/4.0/).

Article

Determining the Composite Structure of Au-Fe-Based Submicrometre Spherical Particles Fabricated by Pulsed-Laser Melting in Liquid

Hokuto Fuse [1], Naoto Koshizaki [1,*], Yoshie Ishikawa [2] and Zaneta Swiatkowska-Warkocka [3]

[1] Graduate School of Engineering, Hokkaido University, Sapporo, Hokkaido 060-8628, Japan; ttx-ken@frontier.hokudai.ac.jp
[2] Nanomaterials Research Institute, National Institute of Advanced Industrial Science and Technology (AIST), Tsukuba, Ibaraki 305-8565, Japan; ishikawa.yoshie@aist.go.jp
[3] Institute of Nuclear Physics, Polish Academy of Sciences, PL-31342 Kraków, Poland; swiatkow@wp.pl
* Correspondence: koshizaki.naoto@eng.hokudai.ac.jp; Tel.: +81-11-706-5594

Received: 4 January 2019; Accepted: 31 January 2019; Published: 3 February 2019

Abstract: Submicrometre spherical particles made of Au and Fe can be fabricated by pulsed-laser melting in liquid (PLML) using a mixture of Au and iron oxide nanoparticles as the raw particles dispersed in ethanol, although the detailed formation mechanism has not yet been clarified. Using a 355 nm pulsed laser to avoid extreme temperature difference between two different raw particles during laser irradiation and an Fe_2O_3 raw nanoparticle colloidal solution as an iron source to promote the aggregation of Au and Fe_2O_3 nanoparticles, we performed intensive characterization of the products and clarified the formation mechanism of Au-Fe composite submicrometre spherical particles. Because of the above two measures (Fe_2O_3 raw nanoparticle and 355 nm pulsed laser), the products—whether the particles are phase-separated or homogeneous alloys—basically follow the phase diagram. In Fe-rich range, the phase-separated Au-core/Fe-shell particles were formed, because quenching induces an earlier solidification of the Fe-rich component as a result of cooling from the surrounding ethanol. If the particle size is small, the quenching rate becomes very rapid and particles were less phase-separated. For high Au contents exceeding 70% in weight, crystalline Au-rich alloys were formed without phase separation. Thus, this aggregation control is required to selectively form homogeneous or phase-separated larger submicrometre-sized particles by PLML.

Keywords: laser melting in liquid; Au-Fe alloy; submicrometre spherical particles; phase separation; reaction control; core-shell particles; laser wavelength; zeta potential

---

1. Introduction

Pulsed-laser melting in liquid (PLML) is a technique derived from pulsed-laser ablation in liquid (PLAL) for nanoparticle fabrication [1,2]. In PLML, raw nanoparticles dispersed in liquid are irradiated by a pulsed-laser with a moderate fluence of about 50–200 mJ pulse^{-1} cm^{-2} (lower fluence than PLAL), resulting in melting and fusion of irradiated particles and subsequently the formation of submicrometre-sized spherical particles via cooling [3–6]. PLML can produce submicrometre spherical particles of various materials, such as metals [7,8], oxides [5,6,9], semiconductors [10,11] and carbides [3,4]. Given the unique features of submicrometre spherical particles—including their dispersibility, stability, crystallinity and sphericity—applications utilizing optical [5,12], medical [13], mechanical [14,15] and magnetic [16] functionality have been examined.

From single-component raw particles, spherical particles are formed without compositional change simply by melting [6]. Reactive fabrication of submicrometre spherical particles with a different composition than the raw particles has been reported for B_4C from B [3,4], Cu from CuO [8] and Fe and

FeO from Fe_3O_4 [17] by a reaction with surrounding organic solvents. Other approaches for forming submicrometre spherical particles of alloys have been intensively investigated from two-component raw particle mixtures, especially for alloy systems of Au and transition metals like Fe, Co and Ni [18–23]. For example, an Au-Co alloy was formed by PLML, although the Au-Co combination was immiscible and could not form an alloy by conventional thermochemical processes [19]. This non-equilibrium process is due to the unique nature of the heating and quenching processes in PLML. In particular, the space-selective pulsed heating of PLML is completely different from conventional heating processes, such as furnace heating. In PLML, the temperature surpasses the melting point of the particles in several hundreds of nanoseconds or shorter, with heating and cooling rates of 10^{11} K s^{-1} and 10^{10} K s^{-1}, respectively [24]. Since pulsed lasers with repetition rate of 10–100 Hz are generally used for PLML, these rapid heating and quenching cycles are repeated many times, with an interval of 10–100 ms for cooling process [25]. Liquid phase surrounding particles acts as a heat dissipation barrier after temporal vaporization and a cooling medium for quenching.

In particular, Au-Fe bimetallic particles have recently been attracting considerable interest because they can be multifunctional materials, combining the plasmonic properties of Au and the magnetic properties of Fe [26–28]. In addition, they may combine to have synergistic functions, such as oxygen evolution, enhanced plasmon absorption, carbon dioxide reduction and imaging and photo-thermal therapy [29–32]. Therefore, various fabrication techniques and formation mechanisms of Au-Fe nanoparticles prepared by PLAL have been intensively investigated [33–37] because the technique's contamination-free nature is beneficial for biological and medical applications.

However, extensive studies on submicrometre-sized Au-Fe particles have not been conducted due to the limited availability of suitable fabrication methods [38]. Our group previously tried to fabricate Au-Fe submicrometre spherical particles by PLML using the second harmonics of an neodymium-doped yttrium aluminium garnet (Nd:YAG) laser at a wavelength of 532 nm [18,22]. Raw particles dispersed in ethanol (C_2H_5OH) were chemically synthesized Fe_3O_4 and laser-ablated Au nanoparticles. During laser irradiation, Fe_3O_4 nanoparticles were reduced by ethanol to FeO or Fe, depending on the laser irradiation conditions. Reduced Fe particles were merged with Au nanoparticles to form submicrometre-sized alloy or composite particles. The heating, reducing, merging, alloying and spheroidizing processes occurred concurrently and submicrometre-sized spherical particles were formed. The magnetic properties of the products were also reported.

Heating behaviour of single particle can be discussed via thermal diffusion length during pulsed laser heating [39]. For 7 ns (pulse width) laser irradiation, thermal diffusion length can be calculated to be 1340 nm for Au fairly larger than the particle size. Thus, even if skin layer of Au particle are selectively heated, the temperature is easily homogenized within Au particle in 7 ns by laser heating. Although the thermal diffusion length of iron oxide is rather difficult to be estimated due to the lack of reliable thermal diffusivity data, it can be roughly estimated to be 350 nm. Thus, the temperature within a single particle can be promptly homogenized during nanosecond laser irradiation.

However, in two-component raw particle systems, the different heating behaviours of each component particle caused by the difference in optical absorption efficiency affects the initial heating step and hence the morphology and inner structure of the products. Figure 1a shows the particle size dependence of the laser fluence required to melt a single particle of Au, Fe_3O_4 and Fe_2O_3 via irradiation with a 532 nm laser light, as calculated based on Mie theory under an adiabatic assumption [40,41]. Although Au has strong optical absorption at this wavelength due to surface plasmon resonance, Fe_3O_4 requires a higher fluence of more than 50 mJ for smaller particles (<100 nm). In contrast, at a wavelength of 355 nm (Figure 1b), the difference in fluence to melt a single particle of each component is less. Thus, the extent of inhomogeneous heating and resultant extreme temperature difference between the respective raw particles is reduced by changing the laser wavelength from 532 nm to 355 nm.

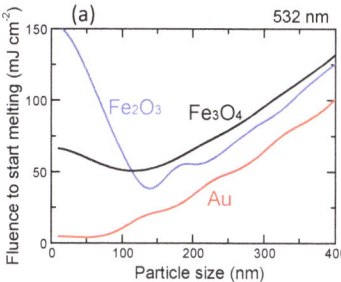 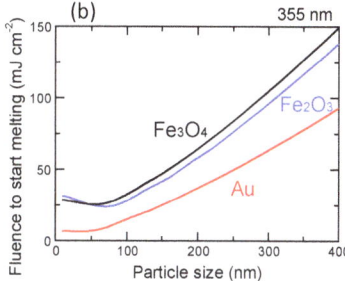

Figure 1. Particle size dependence of laser fluence required to melt a single particle of Au, Fe_3O_4 and Fe_2O_3 calculated based on Mie theory and adiabatic assumption. (**a**) 532 nm. (**b**) 355 nm.

The immediacy of two constituent particles for a prompt reaction during laser irradiation is also an important factor for two-component raw particle systems. Fabricating large, submicrometre spherical particles from raw small nanoparticles during a short laser irradiation time requires such aggregation. In alloy particle formation processes, two different, individually fabricated raw particle solutions have to be mixed. Once dispersed, particles have to encounter one another and combine to form aggregates composed of two different particles. Otherwise, a reaction between the two components would not effectively occur in a short time.

In our previous experiments, chemically produced Fe_3O_4 particles were used as an iron source and the zeta potential of the Fe_3O_4 raw particles fluctuated from -44 to -94 mV in a relatively large negative value. Au nanoparticles fabricated by PLAL are well known to have large negative zeta potentials without any surfactant [42]. Thus, after mixing, these two components are stably dispersed in liquid without a strong interaction between the two raw particles. If positively charged raw particles are available as an iron source, the aggregation proceeds after the raw particles mix because of an electrostatic interaction. Such a source is beneficial for the formation of submicrometre spherical particles based on well-reacted constituent particles.

Here, in contrast to our previous work, a 355 nm laser was adopted for more homogeneous heating of the raw particles and raw Fe_2O_3 particles with positive zeta potential were used for a stronger interaction with negatively charged raw Au nanoparticles. In investigating Au-Fe alloy submicrometre spherical particle fabrication, we especially focused on the detailed concentration-dependence of the products to understand what governs the reaction process in this technique. We also compared our experimental results with those obtained using PLAL for Au-Fe nanoparticle systems. In our discussion, we clarify the procedure to produce particles with thermodynamically stable phases or those with unexpected immiscible alloy phases.

2. Materials and Methods

Raw Au particles were prepared by pulsed-laser ablation of Au plates as a target immersed in ethanol and irradiation with the fundamental wave (1064 nm) of the Nd:YAG laser for 60 min. The average Au particle diameter was estimated to be 10 nm from the relationship between laser fluence and particle size. Raw particles of Fe_2O_3 were purchased from Sigma-Aldrich Japan (Tokyo, Japan, product number: 544884, <50 nm average diameter).

Raw nanoparticles of Au (140 ppm) and Fe_2O_3 (200 ppm) dispersed in ethanol were mixed with different weight ratios of particles in Au:Fe—10%:90%, 20%:80%, ..., 90%:10% in every 10% step. Hereafter, we denote these mixtures as Au10, Au20, ..., Au90, respectively. The third harmonic of the Nd:YAG laser (wavelength: 355 nm, repetition rate: 10 Hz, pulse width: 7 ns) was used to irradiate the mixed solutions (5 mL) with a fluence of 150 mJ pulse^{-1} cm^{-2} for 60 min so that the compositional effect on the resulting products could be studied. The total number of irradiated pulses to the colloidal solution is 3.6×10^4 in this experiment, though the actual irradiated pulses to each particle is around hundredth or thousandth of irradiated pulses due to shadowing effect of laser-absorbing particles and stirring conditions. This laser fluence is sufficiently large to melt Au and Fe_2O_3 raw particles, as well as the produced submicrometre particles (Figure 1b). In addition, to analyse the effect of irradiation time, we irradiated the mixed nanoparticle solution for Au50 for 10, 30, 60 and 120 min.

The obtained submicrometre spherical particles were characterized by a field-emission scanning electron microscope (FE-SEM) (JEOL, JSM-7001FA, Akishima, Japan), transmission electron microscope (TEM) (FEI, Titan3 G2 60-300, Hillsboro, OR, USA) and x-ray diffractometer (XRD) (Rigaku, Ultima IV, Akishima, Japan). The zeta potential of colloidal solutions was measured by a zeta potential/particle size analyser (Beckman Coulter, DelsaNano HC, Brea, CA, USA).

3. Results and Discussion

3.1. Au Concentration Dependence on Internal Structure

Before laser irradiation, the zeta potential of the mixed solution of Au and Fe_2O_3 raw nanoparticles with the different weight ratios was measured. Figure S1 (in Supporting Information) indicates that the absolute values of the zeta potential became closer to zero from those of end members in most of the concentration ranges; hence, particles possibly tend to aggregate or become unstable and the immediacy of two constituent particles was ensured—conditions that are suitable for submicrometre spherical particle fabrication by PLML.

Figure 2 and Figure S2 depict energy-dispersive x-ray spectroscopy (EDS) mapping of Au (red) and Fe (green) and corresponding TEM images of the particles obtained from the raw nanoparticle solution with different mixing ratios. The images in Figure S2 are taken in the high-angle annular dark-field scanning transmission electron microscopy (HAADF-STEM) mode and therefore heavier atoms are brightly indicated, unlike conventional TEM images. Thus, Fe-rich regions (green) in Figure 2 correspond to the dark parts in Figure S2, while Au-rich regions (red) correspond to the bright parts. These figures clearly indicate that the compositional distribution in the particles in Figure 2 corresponded well to the contrast in HAADF-STEM images in Figure S2. Submicrometre spherical particles were formed in all concentration ranges, from Au10 to Au90. From Au10 to Au60, compositional inhomogeneity, mainly in a core (Au-rich)/shell (Fe-rich) structure, was observed. This structure is totally opposite to the core (Fe-rich)/shell (Au-rich)-structured nanoparticles obtained by PLAL [36], which will be discussed later. However, compositionally homogeneous particles were obtained from Au70 to Au90, indicating Au-Fe alloy formation. On the basis of these images, we also found that larger particles tend to be more phase-separated than smaller particles.

Figure 3 depicts the size histograms of the phase-separated particles (black) and homogeneous particles (red) obtained from Figure 2 and Figure S2. Bars placed in the size range between 400–500 nm indicate that the black bar is the frequency for phase-separated particles 400–500 nm in diameter and the red bar is the frequency of homogeneous particles 400–500 nm in diameter. The phase-separated particles (mainly in the core/shell structure) were observed within the Au10–Au60 range, especially in larger particles, whereas homogeneous particles were observed in the smaller size range. In the Au70–Au90 range, all the particles observed were compositionally homogeneous.

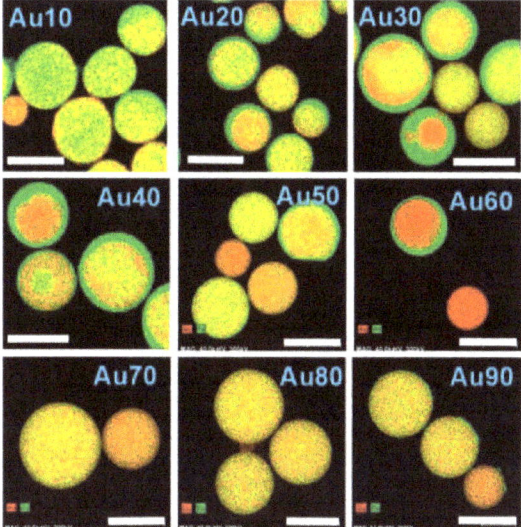

Figure 2. EDS mapping images of particles obtained with different mixing ratios of raw nanoparticle solutions of Au and Fe$_2$O$_3$. "Au10" indicates an Au 10%: Fe 90% in weight. The red and green colours correspond to the Au and Fe components, respectively. Corresponding HAADF-STEM images are shown in Figure S2 in Supporting Information. All scale bars are 500 nm.

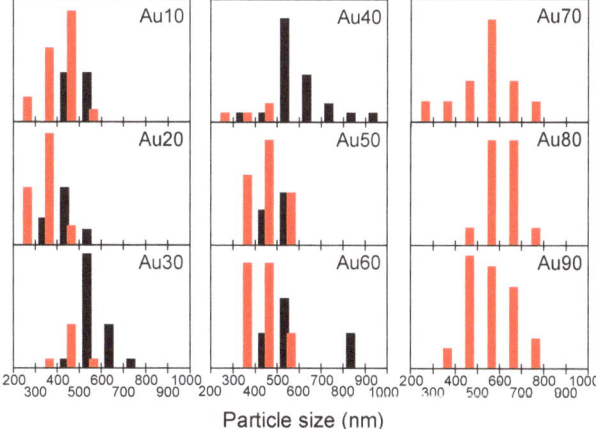

Figure 3. Size histograms of the particles obtained with different mixing ratios of the raw nanoparticle solution of Au and Fe$_2$O$_3$. "Au10" denotes Au 10%:Fe 90% in weight. The black and red bars indicate the frequencies of phase-separated and homogeneous submicrometre particles in every 100 nm step.

Figure 4a shows the homogeneous particle fraction as a function of Au weight percentage. Homogeneous particles for Au30 and Au40 were less abundant. Figure 4b shows the change in average particle diameter of homogeneous and phase-separated submicrometre particles. Even though the relative ratio of phase-separated and homogeneous particles drastically changes with Au weight percentage, as shown in Figure 4a, the average size of both particles gradually increased with the Au content and the size difference between them—about 150 nm—was almost constant across all content ranges.

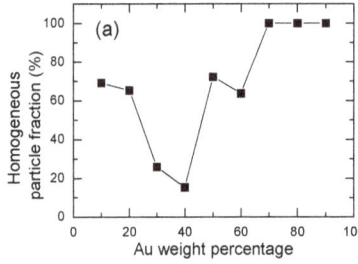
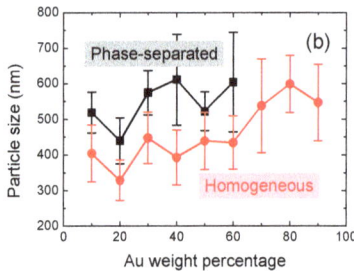

Figure 4. (a) Homogeneous particle fraction as a function of Au weight percentage. (b) Average particle size of phase-separated and homogeneous submicrometre particles as a function of Au weight percentage.

Figure 5 presents XRD patterns of the obtained submicrometre spherical particles. Peaks with open circles indicate face-centred cubic (fcc) AuFe alloys with the end members of fcc-Au and fcc-γ-Fe, which were observed in all mixing ratio ranges. The peak positions of AuFe alloys shifted with the change in Au:Fe mixing ratios. Furthermore, Au peaks from raw particles were also detected from Au10–Au50, suggesting that the phase-separated particles are composed of nearly pure Au metal and Fe-rich Au-Fe alloys in this range. FeO peaks were also observed at the Fe-rich side of Au10 and Au20.

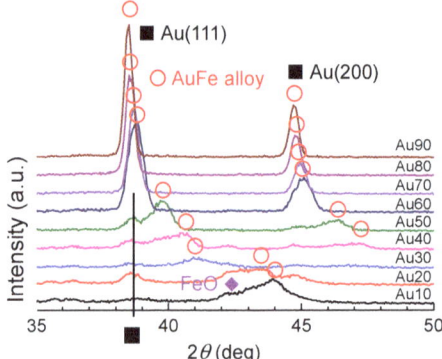

Figure 5. XRD patterns of particles obtained by laser irradiation onto the mixed colloidal solution of Au and Fe_2O_3 in ethanol. "Au10" denotes Au 10%:Fe 90% in weight.

Figure 6 provides the phase diagram of the Au-Fe system redrawn from [43]. On the basis of the horizontal line at 1446 K between Au20 and Au70, the fcc γ-Fe phase (Fe-rich solid solution) is segregated, while Au is still in its melting phase at the initial stage of solidification. Because of the rapid quenching process, phase-separated submicrometre particles composed of fcc Au-rich alloy phases and fcc γ-Fe-rich alloy phases were thus formed.

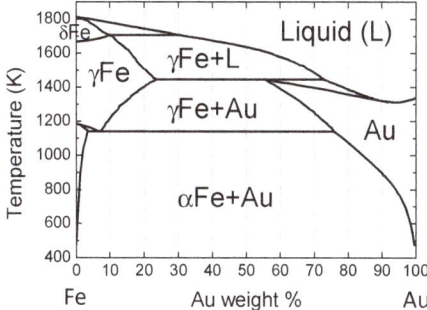

Figure 6. Phase diagram of Au-Fe system.

From the XRD peak shift of the Au-Fe alloy in Figure 5, the composition change of the alloy phase can be estimated, assuming that Vegard's law is true for this system. Figure 7 plots the Au weight percent in the alloy phase, estimated as a function of the weight percent of Au raw particles. The estimated Au content values of produced particles showed a content gap between Au 20 to 70 weight percent, which nearly corresponded to the immiscible gap in Figure 6. However, the generated Au-rich alloy phase has a higher Au content than the feeding composition ratio of the raw particle mixture, suggesting that Fe is removed from the product during the PLML process. This is also supported from yellow colour of supernatant probably from ferric ion during the centrifugation process for XRD sample preparation.

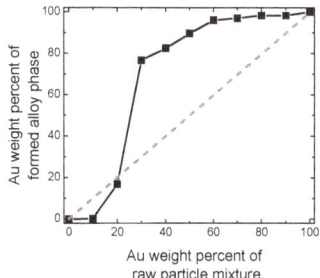

Figure 7. Relationship between the Au weight percentage of the raw particle solution and that of the formed alloy phase after laser irradiation.

3.2. Time Dependence of Produced Particles

Figure 8 and Figure S3 show a plot of the laser irradiation time dependence of elemental mapping and HAADF-STEM images of submicrometre spherical particles obtained by PLML at Au50, respectively. These figures clearly indicate a compositional inhomogeneity in the core (Au-rich)/shell (Fe-rich) structure at short laser irradiation times and a gradual homogenization of the particles by extending the laser irradiation time. The particle size also became larger as irradiation time increased.

Figure 9 shows the XRD patterns of particles obtained by different laser irradiation times onto mixed colloidal solutions of Au50 in ethanol. FeO and Au peaks originating from the raw particles were observed by 10 min laser irradiation, together with the Au-Fe alloy peak. These FeO and Au peaks gradually became smaller by extending the laser irradiation time, implying that the particles were merging and reacting. Peaks from the AuFe alloy (marked as open circles) shifted to pure Au peaks with the increase in laser irradiation time. These results suggest that the Fe component was gradually removed from the alloy system.

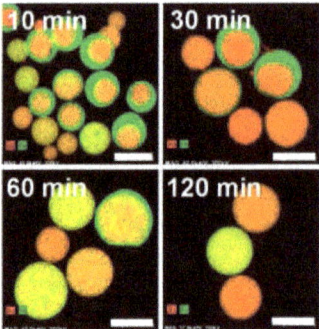

Figure 8. EDS mapping images of particles obtained by different laser irradiation times from the Au50 raw particle solution. Corresponding HAADF-STEM images are shown in Figure S3 in Supporting Information. All scale bars are 500 nm.

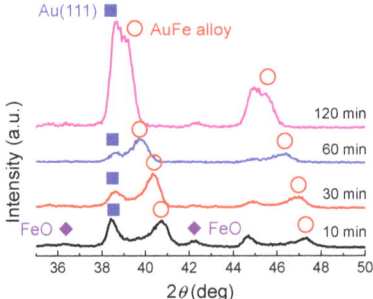

Figure 9. XRD patterns of particles obtained by different laser irradiation time onto the mixed colloidal solution of Au50 in ethanol.

Figure S4 shows the laser irradiation time dependence of the average Au weight percentage of particles measured by EDS. When the irradiation time was 10 min, the Au content was 57.4% in weight, close to the value of the raw particle solution. However, by extending the irradiation time, the percentage of Au increased while that of Fe decreased, indicating a dissolution of the Fe component. This indication is also supported by the colour change of the solution to light yellow, the typical colour of the Fe ion.

3.3. Comparison with Au-Fe Nanoparticles and Formation Mechanism

In the case of Au-Fe nanoparticles obtained by PLAL, Fe-core/Au-shell nanoparticles are usually obtained, in which the core and shell combination is contrary to our case. In the nanoparticle case, the surface area becomes large and the surface energy can dominate the stabilization process of the Fe-core/Au-shell structure because of the low surface energy of Au and short diffusion distance of the nanoparticles [36]. In contrast, in submicrometre particles obtained by PLML, the bulk thermodynamic contribution is dominant, resulting in the solidification of the Fe-rich component at the surface and Au enrichment in the core, as deduced from the phase diagram in Figure 6.

Figure 10 summarizes the formation mechanism of submicrometre spherical particles by PLML. Particles marked in red and in grey are well-aggregated Au and Fe_2O_3 raw nanoparticles. The yellow and green areas correspond to Au-rich and Fe-rich Au-Fe alloys.

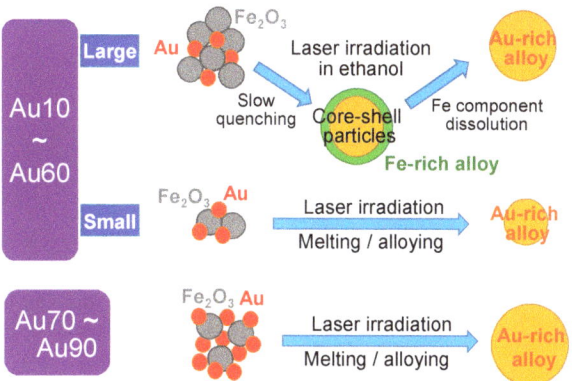

Figure 10. Schematic illustration of the Au-Fe submicrometre spherical particle formation process by PLML in ethanol.

In the case of Au10–Au60 with large particle sizes, the raw particles melt, merge with each other and quench to form alloy particles that are submicrometre in size. However, in this content range on the phase diagram, particles should be phase-separated, with the segregation of Fe-rich AuFe alloy phases at the particle surface because quenching induces an earlier solidification of the Fe-rich component by cooling from the surrounding ethanol. The Au component is pushed toward the centre direction during quenching. The core-shell structure, containing an Au-rich core and Fe-rich shell, is generated by this process. Further laser irradiation induces the thermochemical dissolution of the Fe component to ethanol and the enrichment of Au content in the remaining Au-rich alloy particles.

In the case of Au10–Au60 with small particle sizes, the particles should be phase-separated from the phase diagram, as above. However, the quenching rate is quite fast, owing to the small particle size and particles are solidified before phase separation occurs. Thus, homogeneous particles that might be in an amorphous phase were fabricated. In the case of Au70–Au90, Au-rich homogeneous particles without phase separation were generated, following the phase diagram. When the particle size was small, the particles tended to be single-crystalline by the rapid shifting of the crystallization front.

Previously, our group reported Au-Fe submicrometre spherical particle formation using the 532 nm Nd:YAG laser and chemically fabricated Fe_3O_4 nanoparticles under slightly different laser irradiation conditions (100 mJ, 30 Hz, 60 min) [22]. Raw particles with different relative atomic ratios of Au:Fe—10:1, 1:1, 1:10 (corresponding to Au97, Au78 and Au26 in our notation of the Au-Fe alloy system based on weight percent)—were used as starting materials and the products after laser irradiation were characterized by XRD. At Au97, a nearly Au metal peak was observed, as expected. In contrast, Au metal, broad Au-Fe alloy and FeO peaks were observed at Au78; Au metal, raw Fe_3O_4 and partially reduced FeO peaks were observed at Au26. In both cases, raw and slightly reduced Fe components (FeO) were observed, indicating the insufficient reduction reaction, because these components could not be observed in this experiment. This insufficiency in the previous work might be the effect of a lower fluence, relatively inhomogeneous heating and the less intimacy of raw particles during laser irradiation.

Thus, by adopting a 355 nm laser to enhance homogeneous heating and by selecting raw particles with appropriate surface charge to promote interparticle immediacy, we can predict most of the products by the phase diagram obtained under thermodynamic equilibrium. The only exception in this case was the formation of homogeneous particles smaller than 450 nm for Au 10–Au60 that resulted from the rapid quenching rate of 10^{10} K s^{-1}.

By repetitive laser irradiation, the Au:Fe composition ratio gradually approached the thermodynamically possible alloy composition based on the time-dependence of the products. In contrast, in order to obtain non-equilibrium alloy particles, opposite measures adopted in this study, such as inhomogeneous heating and less encounters between constituent particles, will probably be effective.

4. Conclusions

We studied the reactive fabrication of submicrometre spherical particles combining Au and Fe by PLML using a 355 nm Nd:YAG laser to induce more homogeneous heating of constituent particles and Fe_2O_3 nanoparticles as a raw iron source to promote contact between Au and Fe in liquid. The particles obtained by this process can be explained almost on the basis of the phase diagram whether they are homogeneous or phase-separated. For the content range where Au-Fe phase has to be separated, Fe component is enriched at the surface due to the quenching from the surface for larger particles, while smaller particles tend to be homogeneous particles due to the rapid quenching without crystallization. If we extend laser irradiation time, particles approach the thermodynamically stable compositions. Therefore, in order to fabricate submicrometre alloy particles of immiscible combination, counter measures to this study, inhomogeneous heating and less contact are preferable.

Supplementary Materials: The following are available online at http://www.mdpi.com/2079-4991/9/2/198/s1, Figure S1: Zeta potential of the mixed solution of Au and Fe_2O_3 nanoparticles before laser irradiation, Figure S2: HAADF-STEM images of particles obtained with different mixing ratios of raw nanoparticle solutions of Au and Fe_2O_3, Figure S3: HAADF-STEM images of particles obtained by different laser irradiation times from the Au50 raw particle solution, Figure S4: Au weight percent change in the produced submicrometre particles with the laser irradiation time for Au50.

Author Contributions: Conceptualization, N.K., Y.I. and Z.S.-W.; methodology, H.F., N.K.; validation, H.F., N.K. and Y.I.; formal analysis, H.F., N.K.; investigation, H.F., N.K.; resources, N.K.; data curation, H.F.; writing—original draft preparation, N.K. Y.I. and Z.S.-W.; writing—review and editing, N.K., Y.I. and Z.S.-W.; visualization, N.K.; supervision, N.K.; project administration, N.K.; funding acquisition, N.K.

Funding: This research was partially funded by JSPS KAKENHI Grant Numbers 26289266 and 26870908.

Acknowledgments: This work was conducted at "Joint-use Facilities: Laboratory of Nano-Micro Material Analysis", Hokkaido University, supported by "Nanotechnology Platform" Program of the Ministry of Education, Culture, Sports, Science and Technology (MEXT), Japan.

Conflicts of Interest: The authors declare no conflict of interest.

References

1. Zhang, D.; Gökce, B.; Barcikowski, S. Laser Synthesis and Processing of Colloids: Fundamentals and Applications. *Chem. Rev.* **2017**, *117*, 3990–4103. [CrossRef] [PubMed]
2. Xiao, J.; Liu, P.; Wang, C.X.; Yang, G.W. External field-assisted laser ablation in liquid: An efficient strategy for nanocrystal synthesis and nanostructure assembly. *Prog. Mater. Sci.* **2017**, *87*, 140–220. [CrossRef]
3. Ishikawa, Y.; Shimizu, Y.; Sasaki, T.; Koshizaki, N. Boron carbide spherical particles encapsulated in graphite prepared by pulsed laser irradiation of boron in liquid medium. *Appl. Phys. Lett.* **2007**, *91*, 161110. [CrossRef]
4. Ishikawa, Y.; Feng, Q.; Koshizaki, N. Growth fusion of submicron spherical boron carbide particles by repetitive pulsed laser irradiation in liquid media. *Appl. Phys. A* **2010**, *99*, 797–803. [CrossRef]
5. Wang, H.; Miyauchi, M.; Ishikawa, Y.; Pyatenko, A.; Koshizaki, N.; Li, Y.; Li, L.; Li, X.; Bando, Y.; Golberg, D. Single-Crystalline Rutile TiO_2 Hollow Spheres: Room-Temperature Synthesis, Tailored Visible-Light-Extinction, and Effective Scattering Layer for Quantum Dot-Sensitized Solar Cells. *J. Am. Chem. Soc.* **2011**, *133*, 19102–19109. [CrossRef] [PubMed]
6. Wang, H.; Koshizaki, N.; Li, L.; Jia, L.; Kawaguchi, K.; Li, X.; Pyatenko, A.; Swiatkowska-Warkocka, Z.; Bando, Y.; Golberg, D. Size-Tailored ZnO Submicrometer Spheres: Bottom-Up Construction, Size-Related Optical Extinction, and Selective Aniline Trapping. *Adv. Mater.* **2011**, *23*, 1865–1870. [CrossRef] [PubMed]
7. Tsuji, T.; Yahata, T.; Yasutomo, M.; Igawa, K.; Tsuji, M.; Ishikawa, Y.; Koshizaki, N. Preparation and investigation of the formation mechanism of submicron-sized spherical particles of gold using laser ablation and laser irradiation in liquids. *Phys. Chem. Chem. Phys.* **2013**, *15*, 3099–3107. [CrossRef] [PubMed]

8. Wang, H.; Pyatenko, A.; Kawaguchi, K.; Li, X.; Swiatkowska-Warkocka, Z.; Koshizaki, N. Selective Pulsed Heating for the Synthesis of Semiconductor and Metal Submicrometer Spheres. *Angew. Chem. Inter. Ed.* **2010**, *49*, 6361–6364. [CrossRef] [PubMed]
9. Ishikawa, Y.; Koshizaki, N.; Pyatenko, A.; Saitoh, N.; Yoshizawa, N.; Shimizu, Y. Nano- and Submicrometer-Sized Spherical Particle Fabrication Using a Submicroscopic Droplet Formed Using Selective Laser Heating. *J. Phys. Chem. C* **2016**, *120*, 2439–2446. [CrossRef]
10. Li, X.; Pyatenko, A.; Shimizu, Y.; Wang, H.; Koga, K.; Koshizaki, N. Fabrication of Crystalline Silicon Spheres by Selective Laser Heating in Liquid Medium. *Langmuir* **2011**, *27*, 5076–5080. [CrossRef]
11. Wang, H.; Li, X.; Pyatenko, A.; Koshizaki, N. Gallium Phosphide Spherical Particles by Pulsed Laser Irradiation in Liquid. *Sci. Adv. Mater.* **2012**, *4*, 544–547. [CrossRef]
12. Fujiwara, H.; Niyuki, R.; Ishikawa, Y.; Koshizaki, N.; Tsuji, T.; Sasaki, K. Low-threshold and quasi-single-mode random laser within a submicrometer-sized ZnO spherical particle film. *Appl. Phys. Lett.* **2013**, *102*, 061110. [CrossRef]
13. Iwagami, T.; Ishikawa, Y.; Koshizaki, N.; Yamamoto, N.; Tanaka, H.; Masunaga, S.; Sakurai, Y.; Kato, I.; Iwai, S.; Suzuki, M.; et al. Boron Carbide Particle as a Boron Compound for Boron Neutron Capture Therapy. *J. Nucl. Med. Radiat. Ther.* **2014**, *5*, 177. [CrossRef]
14. Hu, X.; Gong, H.; Wang, Y.; Chen, Q.; Zhang, J.; Zheng, S.; Yang, S.; Cao, B. Laser-induced reshaping of particles aiming at energy-saving applications. *J. Mater. Chem.* **2012**, *22*, 15947–15952. [CrossRef]
15. Kondo, M.; Shishido, N.; Kamiya, S.; Kubo, A.; Umeno, Y.; Ishikawa, Y.; Koshizaki, N. High-Strength Sub-Micrometer Spherical Particles Fabricated by Pulsed Laser Melting in Liquid. *Part. Part. Syst. Charact.* **2018**, *35*, 1800061. [CrossRef]
16. Swiatkowska-Warkocka, Z.; Kawaguchi, K.; Wang, H.; Katou, Y.; Koshizaki, N. Controlling exchange bias in Fe_3O_4/FeO composite particles prepared by pulsed laser irradiation. *Nanoscale Res. Lett.* **2011**, *6*, 226. [CrossRef] [PubMed]
17. Ishikawa, Y.; Koshizaki, N.; Pyatenko, A. Submicrometer-Sized Spherical Iron Oxide Particles Fabricated by Pulsed Laser Melting in Liquid. *Electron. Commun. Jpn.* **2016**, *99*, 37–42. [CrossRef]
18. Swiatkowska-Warkocka, Z.; Kawaguchi, K.; Shimizu, Y.; Pyatenko, A.; Wang, H.; Koshizaki, N. Synthesis of Au-Based Porous Magnetic Spheres by Selective Laser Heating in Liquid. *Langmuir* **2012**, *28*, 4903–4907. [CrossRef] [PubMed]
19. Swiatkowska-Warkocka, Z.; Koga, K.; Kawaguchi, K.; Wang, H.; Pyatenko, A.; Koshizaki, N. Pulsed laser irradiation of colloidal nanoparticles: A new synthesis route for the production of non-equilibrium bimetallic alloy submicrometer spheres. *RSC Adv.* **2013**, *3*, 79–83. [CrossRef]
20. Swiatkowska-Warkocka, Z.; Pyatenko, A.; Krok, F.; Jany, B.R.; Marszalek, M. Synthesis of new metastable nanoalloys of immiscible metals with a pulse laser technique. *Sci. Rep.* **2015**, *5*, 9849. [CrossRef]
21. Swiatkowska-Warkocka, Z.; Pyatenko, A.; Koshizaki, N.; Kawaguchi, K. Synthesis of various 3D porous gold-based alloy nanostructures with branched shapes. *J. Colloid Interface Sci.* **2016**, *483*, 281–286. [CrossRef] [PubMed]
22. Swiatkowska-Warkocka, Z.; Pyatenko, A.; Koga, K.; Kawaguchi, K.; Wang, H.; Koshizaki, N. Various Morphologies/Phases of Gold-Based Nanocomposite Particles Produced by Pulsed Laser Irradiation in Liquid Media: Insight in Physical Processes Involved in Particles Formation. *J. Phys. Chem. C* **2017**, *121*, 8177–8187. [CrossRef]
23. Swiatkowska-Warkocka, Z.; Pyatenko, A.; Shimizu, Y.; Perzanowski, M.; Zarzycki, A.; Jany, R.B.; Marszalek, M. Tailoring of Magnetic Properties of NiO/Ni Composite Particles Fabricated by Pulsed Laser Irradiation. *Nanomaterials* **2018**, *8*, 790. [CrossRef] [PubMed]
24. Sakaki, S.; Ikenoue, H.; Tsuji, T.; Ishikawa, Y.; Koshizaki, N. Pulse-Width Dependence of the Cooling Effect on Sub-Micrometer ZnO Spherical Particle Formation by Pulsed-Laser Melting in a Liquid. *ChemPhysChem* **2017**, *18*, 1101–1107. [CrossRef] [PubMed]
25. Sakaki, S.; Ikenoue, H.; Tsuji, T.; Ishikawa, Y.; Koshizaki, N. Influence of pulse frequency on synthesis of nano and submicrometer spherical particles by pulsed laser melting in liquid. *Appl. Surf. Sci.* **2018**, *435*, 529–534. [CrossRef]
26. Amendola, V.; Meneghetti, M.; Bakr, O.M.; Riello, P.; Polizzi, S.; Anjum, D.H.; Fiameni, S.; Arosio, P.; Orlando, T.; de Julian Fernandez, C.; et al. Coexistence of plasmonic and magnetic properties in Au89Fe11 nanoalloys. *Nanoscale* **2013**, *5*, 5611–5619. [CrossRef] [PubMed]

27. Amendola, V.; Scaramuzza, S.; Agnoli, S.; Polizzi, S.; Meneghetti, M. Strong dependence of surface plasmon resonance and surface enhanced Raman scattering on the composition of Au–Fe nanoalloys. *Nanoscale* **2014**, *6*, 1423–1433. [CrossRef] [PubMed]
28. Amendola, V.; Scaramuzza, S.; Litti, L.; Meneghetti, M.; Zuccolotto, G.; Rosato, A.; Nicolato, E.; Marzola, P.; Fracasso, G.; Anselmi, C.; et al. Magneto-Plasmonic Au-Fe Alloy Nanoparticles Designed for Multimodal SERS-MRI-CT Imaging. *Small* **2014**, *10*, 2476–2486. [CrossRef] [PubMed]
29. Vassalini, I.; Borgese, L.; Mariz, M.; Polizzi, S.; Aquilanti, G.; Ghigna, P.; Sartorel, A.; Amendola, V.; Alessandri, I. Enhanced Electrocatalytic Oxygen Evolution in Au–Fe Nanoalloys. *Angew. Chem. Inter. Ed.* **2017**, *56*, 6589–6593. [CrossRef] [PubMed]
30. Amendola, V.; Saija, R.; Maragò, O.M.; Iatì, M.A. Superior plasmon absorption in iron-doped gold nanoparticles. *Nanoscale* **2015**, *7*, 8782–8792. [CrossRef]
31. Sun, K.; Cheng, T.; Wu, L.; Hu, Y.; Zhou, J.; Maclennan, A.; Jiang, Z.; Gao, Y.; Goddard, W.A.; Wang, Z. Ultrahigh Mass Activity for Carbon Dioxide Reduction Enabled by Gold–Iron Core–Shell Nanoparticles. *J. Am. Chem. Soc.* **2017**, *139*, 15608–15611. [CrossRef] [PubMed]
32. Timothy, A.L.; James, B.; Jesse, A.; Konstantin, S. Hybrid plasmonic magnetic nanoparticles as molecular specific agents for MRI/optical imaging and photothermal therapy of cancer cells. *Nanotechnology* **2007**, *18*, 325101.
33. Scaramuzza, S.; Agnoli, S.; Amendola, V. Metastable alloy nanoparticles, metal-oxide nanocrescents and nanoshells generated by laser ablation in liquid solution: Influence of the chemical environment on structure and composition. *Phys. Chem. Chem. Phys.* **2015**, *17*, 28076–28087. [CrossRef] [PubMed]
34. Amendola, V.; Scaramuzza, S.; Carraro, F.; Cattaruzza, E. Formation of alloy nanoparticles by laser ablation of Au/Fe multilayer films in liquid environment. *J. Colloid Interface Sci.* **2017**, *489*, 18–27. [CrossRef] [PubMed]
35. Wagener, P.; Jakobi, J.; Rehbock, C.; Chakravadhanula, V.S.K.; Thede, C.; Wiedwald, U.; Bartsch, M.; Kienle, L.; Barcikowski, S. Solvent-surface interactions control the phase structure in laser-generated iron-gold core-shell nanoparticles. *Sci. Rep.* **2016**, *6*, 23352. [CrossRef] [PubMed]
36. Tymoczko, A.; Kamp, M.; Prymak, O.; Rehbock, C.; Jakobi, J.; Schürmann, U.; Kienle, L.; Barcikowski, S. How the crystal structure and phase segregation of Au–Fe alloy nanoparticles are ruled by the molar fraction and size. *Nanoscale* **2018**, *10*, 16434–16437. [CrossRef]
37. Kamp, M.; Tymoczko, A.; Schürmann, U.; Jakobi, J.; Rehbock, C.; Rätzke, K.; Barcikowski, S.; Kienle, L. Temperature-Dependent Ultrastructure Transformation of Au–Fe Nanoparticles Investigated by in Situ Scanning Transmission Electron Microscopy. *Cryst. Growth Des.* **2018**, *18*, 5434–5440. [CrossRef]
38. Majerič, P.; Jenko, D.; Friedrich, B.; Rudolf, R. Formation of Bimetallic Fe/Au Submicron Particles with Ultrasonic Spray Pyrolysis. *Metals* **2018**, *8*, 278. [CrossRef]
39. Sakaki, S.; Saitow, K.; Sakamoto, M.; Wada, H.; Swiatkowska-Warkocka, Z.; Ishikawa, Y.; Koshizaki, N. Comparison of picosecond and nanosecond lasers for the synthesis of TiN sub-micrometer spherical particles by pulsed laser melting in liquid. *Appl. Phys. Exp.* **2018**, *11*, 035001. [CrossRef]
40. Pyatenko, A.; Wang, H.; Koshizaki, N.; Tsuji, T. Mechanism of pulse laser interaction with colloidal nanoparticles. *Laser Photon. Rev.* **2013**, *7*, 596–604. [CrossRef]
41. Pyatenko, A.; Wang, H.; Koshizaki, N. Growth Mechanism of Monodisperse Spherical Particles under Nanosecond Pulsed Laser Irradiation. *J. Phys. Chem. C* **2014**, *118*, 4495–4500. [CrossRef]
42. Palazzo, G.; Valenza, G.; Dell'Aglio, M.; De Giacomo, A. On the stability of gold nanoparticles synthesized by laser ablation in liquids. *J. Colloid Interface Sci.* **2017**, *489*, 47–56. [CrossRef] [PubMed]
43. Cahn, R.W. *Binary Alloy Phase Diagrams*, 2nd ed.; Massalski, T.B., Okamoto, H., Subramanian, P.R., Kacprzak, L., Eds.; ASM International: Materials Park, OH, USA, 1990.

© 2019 by the authors. Licensee MDPI, Basel, Switzerland. This article is an open access article distributed under the terms and conditions of the Creative Commons Attribution (CC BY) license (http://creativecommons.org/licenses/by/4.0/).

Article

A Controllability Investigation of Magnetic Properties for FePt Alloy Nanocomposite Thin Films

Jian Yu [1], Tingting Xiao [1], Xuemin Wang [1], Xiuwen Zhou [1], Xinming Wang [1], Liping Peng [1], Yan Zhao [1], Jin Wang [2], Jie Chen [1], Hongbu Yin [1] and Weidong Wu [1,3,4,*]

[1] Science and Technology on Plasma Physics Laboratory, Research Center of Laser fusion, China Academy of Engineering Physics, Mianyang 621900, China; yujianroy@163.com (J.Y.); tingtingxiao@yeah.net (T.X.); wangxuemin75@sina.com (X.W.); xiuwenzhou@caep.cn (X.Z.); 3965@163.com (X.W.); pengliping2005@126.com (L.P.); zhaoyan267@163.com (Y.Z.); chenjie1067@163.com (J.C.); yhp1214@mail.ustc.edu.cn (H.Y.)
[2] State Key Laboratory of Advanced Technology for Materials Synthesis and Processing, Wuhan University of Technology, Wuhan 430070, China; swustwj@163.com
[3] School of Materials Science and Engineering, Southwest University of Science and Technology, Mianyang 621000, China
[4] Collaborative Innovation Center of IFSA (CICIFSA), Shanghai Jiao Tong University, Shanghai 200240, China
* Correspondence: wuweidongding@163.com

Received: 5 December 2018; Accepted: 27 December 2018; Published: 3 January 2019

Abstract: An appropriate writing field is very important for magnetic storage application of $L1_0$ FePt nanocomposite thin films. However, the applications of pure $L1_0$ FePt are limited due to its large coercivity. In this paper, the ratios of $L1_0$ and non-$L1_0$ phase FePt alloy nanoparticles in FePt/MgO (100) nanocomposite thin films were successfully tuned by pulsed laser deposition method. By adjusting the pulsed laser energy density from 3 to 7 J/cm^2, the ordering parameter initially increased, and then decreased. The highest ordering parameter of 0.9 was obtained at the pulsed laser energy density of 5 J/cm^2. At this maximum value, the sample had the least amount of the soft magnetic phase of almost 0%, as analyzed by a magnetic susceptibility study. The saturation magnetization decreased with the increase in the content of soft magnetic phase. Therefore, the magnetic properties of FePt nanocomposite thin films can be controlled, which would be beneficial for the magnetic applications of these thin films.

Keywords: pulse laser deposition; FePt alloy; magnetic phase

---

1. Introduction

The face-centered-tetragonal (fct) $L1_0$-FePt alloy with large magnetocrystalline anisotropy content was regarded as the most promising material for ultra high density perpendicular magnetic recording [1,2]. However, the high coercivity of FePt greatly exceeds the writing field of available heads, which is limited by the head materials [3]. Thus, it is necessary to find a way to reduce the writing field. Exchange coupling between hard magnetic and soft magnetic phase has been proposed to solve this problem. In order to realize the exchange coupling in materials, two or more phases are necessary in the composite [4–6].

Furthermore, Fe-rich Fe_3Pt, Pt-rich $FePt_3$ or even disordered face-centered cubic (fcc) alloys can exhibit valuable magnetic properties [7–9]. Therefore, combing these materials with $L1_0$ FePt has attracted the attention of researchers in the past few years [10–13]. Sun's group [11] used the reduction of platinum acetylacetonate and decomposition of iron pentacarbonyl in the presence of oleic acid and oleyl amine stabilizers to synthesize FePt alloy with different Pt concentrations, and studied the relationship between coercivity and Pt concentration. Lin [12] fabricated nanocomposite

FePt-FePt$_3$ films by annealing the (Pt/Fe)$_{10}$ multilayer film, and focused on the influence of annealing temperature. Suber [13] systemically studied interactions between hard and soft magnetic phases by thermal treatment of core-shell FePt (Ag)@Fe$_3$O$_4$. However, most of the methods to prepare FePt exchange coupling materials involve chemical synthesis, which is expensive and cannot precisely control the proportion of the components. Moreover, the annealing process, which is required for the formation of L1$_0$ phase, will lead to undesired particle agglomeration, giving rise to clusters of individual particles [14–16]. Since pulse laser deposition (PLD) has the advantages of fast growth rate and easily adjustable process parameters, the prepared samples are free of impurities [17]. Therefore, PLD can be developed as a way to fabricate FePt alloy with different component proportions. Furthermore, compared with other chemical synthesis methods, it is easy to embed FePt nanoparticles in a matrix by PLD method, which can effectively prevent the agglomeration of FePt particles [18].

In this work, FePt nanoparticles with different proportions of Pt and Fe were embedded in epitaxial MgO by a PLD method. The ratio of Pt and Fe was confirmed by X-ray photoelectron spectroscopy (XPS). X-ray diffractometer (XRD), and high resolution transmission electron microscope (HRTEM) equipped with an energy dispersive X-ray detector (EDX) were used to analyze the structure of samples with different components. The magnetic properties were measured by superconducting quantum interference device (SQUID). The influence of the pulsed laser energy density of PLD on the structure and magnetic properties of samples was studied.

2. Experimental

FePt/MgO nanocomposite films were grown on MgO (100) substrate by pulsed laser deposition (PLD), which equipped a KrF excimer laser (Anhui Institute of Optics and Fine Mechanics (Hefei, China)) with a wavelength of 248 nm and pulse width of 25 ns. The pure MgO target (purity 99.99%) and an alloy target with a composition of Fe$_{50}$Pt$_{50}$ were used to fabricate the FePt/MgO nanocomposite films under an ultrahigh-vacuum (UHV) system. Before the fabrication of samples, the substrate was heated and degassed for 3 h at 1033 K. The MgO buffer layer was subsequently epitaxially grown on the substrate. Afterwards, the FePt layer of about 6 nm was grown on MgO buffer layer. Subsequently, MgO protective layers were deposited onto the sample. The sample was annealed for about 20 min after the deposition of every MgO layer. The pulsed laser energy density for the growth of MgO was 4 J/cm^2. For fabricating FePt, the laser energy density was varied from 3, 4, 5, 6 and 7 J/cm^2, and the samples were labelled as samples 1$^\#$, 2$^\#$, 3$^\#$, 4$^\#$ and 5$^\#$, respectively. Finally, all of the as-deposited FePt/MgO nanocomposite films were post-annealed at 1173 K for 4 h under a Ar + H$_2$ (5%) flowing gas atmosphere. The experimental parameters for fabricating the FePt/MgO nanocomposite films are listed in Table 1.

Table 1. Experimental parameters for FePt/MgO nanocomposite film fabrication.

Experiment Conditions	Experiment Parameters
background vacuum	3.0×10^{-6} Pa
working vacuum	5.0×10^{-5} Pa
target	MgO purity > 99.99% FePt purity > 99.99%
substrate	MgO (100)
laser pulse frequency	2 Hz
number of pulses	MgO: 1200 pulses
distance between the target and substrate	5 cm

XPS analysis (Mg Kα, 1253.6 eV, ThermoFisher Scientific (Waltham, MA, USA)) and EDX (JEOL, Tokyo, Japan) was employed to determine the chemical composition and the proportions of Fe and Pt, and the XPS spectra were fitted using the XPSPEAK41 program and Shirley-type background. The phase composition of the films were identified using XRD (θ–2θ (symmetric reflection) diffraction geometry, Rigaku, Tokyo, Japan) with CuKα radiation of 1.5418 Å wavelength and HRTEM (JEOL,

Tokyo, Japan). The magnetic properties at room temperature were measured by a SQUID (Quantum Design (SanDiego, CA, USA)) in the range of −5 to 5 T.

3. Results and Discussion

XPS measurement was used to confirm the chemical composition and Pt/Fe ratios for the different samples. It can be seen from the XPS survey spectra of the samples (Figure 1a) that the elements iron, platinum, magnesium and oxygen were present in all the samples. There was no impurity element in FePt/MgO nanocomposite films. Thus, the stability of structure and properties of the samples were ensured. As shown in Figure 1b,d, six peaks are required to fit Fe 2p spectra of sample 1# and sample 5#, respectively. The peak A and D near 707.2 ± 0.2 eV and 720.4 ± 0.2 eV correspond to $2p_{3/2}$ and $2p_{1/2}$ of pure Fe, respectively. Our previous work has confirmed that there is a peak shift toward to higher binding energy due to the bonding of the Fe and Pt in the single unit [19]. Therefore, the peak B and E located at 710.1 ± 0.2 eV and 724.0±0.2 eV correspond to $2p_{3/2}$ and $2p_{1/2}$ of pure Fe of ordered FePt alloy, respectively. And it can be confirm that the Peak A and D are contributed by the Fe of disordered FePt alloy. Besides, there are two small peaks C and F in the detailed spectra of Fe 2p with a binding energy at 714.9 ± 0.2 eV and 729.8 ± 0.2 eV, respectively. They are assigned to Fe in Fe_3O_4. This may be caused by the diffusion of oxygen from MgO onto the surface of FePt nanoparticles. The Pt 4f peaks located at 71.50 ± 0.2 and 74.80 ± 0.2 eV corresponding to the Pt $4f_{7/2}$ and Pt $4f_{5/2}$ can be identified in Figure 1c,e, respectively. These results confirm that the valence states of Fe and Pt were not affected by the pulsed laser energy density.

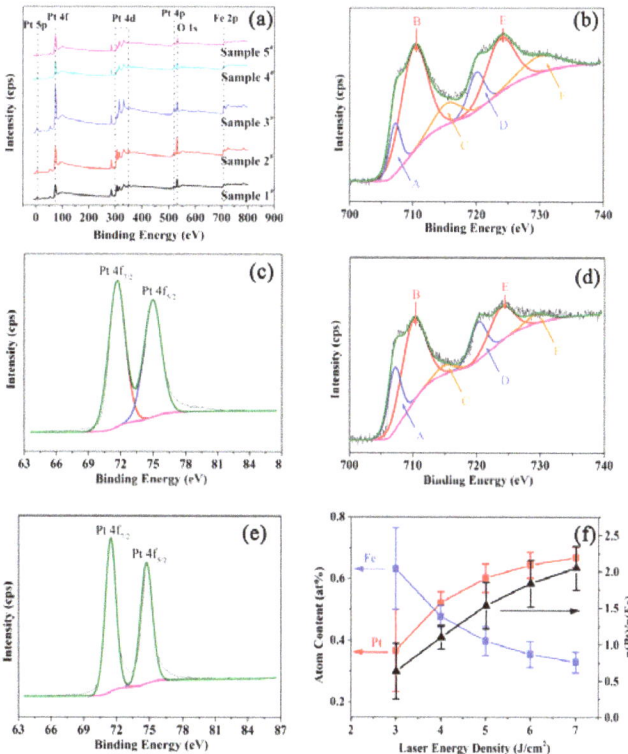

Figure 1. (a) XPS survey spectrum for FePt/MgO nanocomposite film, XPS spectrum for Fe 2p (b,d) and Pt 4f (c,e) for sample 1# and sample 5#, respectively. (f) atom content for Fe and Pt, and the Pt/Fe atom ratio for different samples calculated by XPS.

To investigate the molar composition of FePt/MgO composite film, the equations follows was used to calculate the ratio of Fe to Pt with the areas under peaks of elements in the spectra [20].

$$\frac{n_{Fe}}{n_{Pt}} = \frac{S_{Fe}/g_{Fe}}{S_{Pt}/g_{Pt}} \tag{1}$$

where S is the area under the peak and g represents the atomic sensitivity factor. The g value was set as 10.54 and 1.54 for Fe and Pt, respectively. Figure 1f plots the results as a function of pulsed laser energy density. The ratio of Fe to Pt for FePt/MgO nanocomposite films was calculated with the areas under peaks of elements in the spectra. The Fe atom percentages for sample $1^{\#}$ to sample $5^{\#}$ were 67.46, 48.34, 42.38, 40.50 and 36.56 at%, respectively. Moreover, the Pt/Fe atom ratios for the five samples were calculated as 0.48, 1.07, 1.36, 1.47 and 1.73, respectively. It can be concluded that the ratio of Pt to Fe increased with the increase in the pulsed laser energy density. The proportion of components could be controlled between $Fe_{67}Pt_{33}$ and $Fe_{37}Pt_{63}$. This phenomenon can be explained as follows. As iron and platinum have different melting points, the element contents evaporated from the target surface were different during the interaction between laser and target. The plasma plume generated by the laser ablation target thereby had different element contents. Therefore, the ratio of Pt to Fe varied with different pulsed laser energy densities.

In order to confirm the structure of the FePt layer, XRD measurement were performed, and the results are shown in Figure 2. The peaks were labelled using lattice parameters for the bulk tetragonal FePt (a = 0.3847nm, c = 0.3715 nm, P4/mmm, see PDF#43-1359). It can be seen from Figure 2a that the superlattice diffraction peaks (001) and (002) can be distinguished in the patterns for all samples, which indicates that the FePt in all samples was arranged in an ordered tetragonal $L1_0$ phase, and that the films had high single-orientation of c-axis. It can be clearly seen from the expanded view in Figure 2b that with the increase in the pulsed laser energy density, the characteristic diffraction peak (001) of FePt shifted to lower angle. According to the Bragg's formula

$$2d \sin \theta = n\lambda \tag{2}$$

where the d, θ and λ are the interplanar crystal spacing, angle between the incident beam and the crystal face and X-ray wavelength, respectively. The shifting to the lower angle indicates the increase in the lattice parameter. This can be explained by Vegard's law

$$a_{A_{(1-x)}B_x} = (1-x)a_A + xa_B \tag{3}$$

where $a_{A(1-x)Bx}$ is the lattice parameter of the solution, a_A and a_B are the lattice parameters of the molar fraction of B in the solution. It was found from the XPS analysis that the Pt/Fe ratio increased with the raise of the pulsed laser energy density, and the lattice parameter for Pt (0.39242 nm) was larger than that of Fe (0.28664 nm). The Pt atom occupied the Fe atom position when the Pt/Fe ratio increased. Therefore, the lattice parameter increases, and the characteristic diffraction peak (001) of FePt shifted to a lower angle for samples with the increase in the pulsed laser energy density.

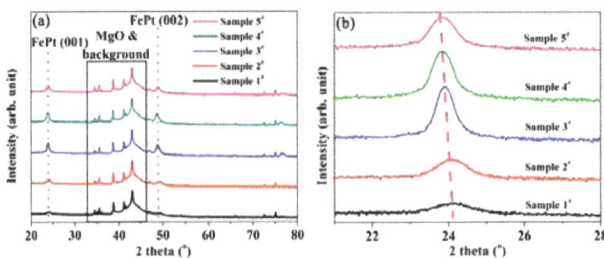

Figure 2. (a) The XRD patterns of FePt/MgO nanocomposite films; (b) an expanded view of peak (001) of FePt.

To further investigate the growth degree of $L1_0$-FePt, the ordering parameter was evaluated as shown below [21]:

$$S = \left[\left(\frac{I_{(001)}}{I_{(002)}}\right) \times \left(\frac{F_f}{F_s}\right)^2 \frac{(L \times A \times D)_f}{(L \times A \times D)_s}\right]^{1/2} = k \times \left(\frac{I_{(001)}}{I_{(002)}}\right)^{1/2} \quad (4)$$

where $I_{(001)}$ and $I_{(002)}$ are the peak intensities of FePt (001) and FePt (002); F, L, A, and D refer to the structure factor, Lorentz polarization factor, absorption factor, and temperature factor, respectively. The terms f and s represent the fundamental peak and superlattice peak, respectively. The k value (0.59) is obtained by reference [22]. Figure 3 presents the plot of ordering parameter S as a function of pulsed laser energy density. With the increase in the laser energy density, the ordering parameter increased. When the laser energy density was 5 J/cm² for sample 3#, and the ordering parameter reached a maximum of 0.90 and then decreased. The reason for this phenomenon can be explained as follows. The excess Pt over the stoichiometric composition could help with atomic diffusion while too much content would affect the structure of $L1_0$ order phase [23].

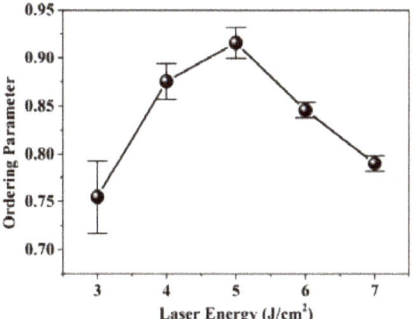

Figure 3. The ordering parameter and FWHM of FePt (001) peak.

Figure 4 shows TEM images of the FePt/MgO nanocomposite films. It can be seen from Figure 4a that the FePt nanoparticles were well-separated in the MgO matrix, which indicated that the nucleation regime of FePt in samples was Volmer-Weber-like growth [24]. Figure 4b–d are the HRTEM images of samples 1#–3#, respectively. It can be seen from the images that changing the pulsed laser energy density did not influence the morphology of FePt/MgO nanocomposite films. The FePt nanoparticles remained spherical and embedded in the MgO matrix. Moreover the superstructure became evident from the alternating bright and dark contrast of the lattice planes. This was due to the largely different electronic scattering cross sections for Fe and Pt [25]. However, with the increase in the pulsed laser energy density, the lattice parameter of the FePt nanoparticles changed. For sample 3# shown in Figure 4c, the lattice parameter was 0.386 nm, which indicates the existence of $L1_0$ FePt. And the EDX results for sample 3# (shown in Figure S1 and Table S1) indicate that the Fe/Pt ratio is about 50:50. This result is consistent with the XPS analysis. For samples 1# and 5#, as shown in Figure 4b,d, the lattice parameterswere0.348 and 0.403 nm, respectively. It indicates that the lattice parameter of the FePt alloy increased with the increase in pulsed laser energy density. This result was consistent with the XRD analysis.

Magnetic properties of FePt/MgO nanocomposite films fabricated by different pulsed laser energy densities were measured by SQUID. Figure 5 shows the out-plane hysteresis loops of all samples. As seen from Figure 5, all the samples showed strong ferromagnetic properties. At low magnetic field there was an obvious weak kink when the hysteresis passed through the magnetization axis for samples 1#, 4# and 5#. However, the hysteresis loops of samples 2# and 3# showed almost smooth curves at low magnetic field. This phenomenon suggests that there was a composite phase consisting of a hard magnetic phase ($L1_0$) and soft magnetic phase (A1 or $L1_2$) [26]. The reason for the appearance of kinks in the hysteresis curve was possibly the exchange coupling between the hard and soft magnetic phases [27,28]. In general, for mixed magnetic materials, increasing the percentage of soft magnetic phase will enlarge the kink. Therefore the decrease in magnetization at the kink can be used to estimate the fraction of soft magnetic phase in the mixture. Combined with the XPS result, it was evident that the different samples had different Pt/Fe ratios. For the Fe-rich sample 1#, the amount of soft phasewas about 47.5%, which may contain FePt fcc phase or Fe_3Pt, while samples 4# and 5# probably contained FePt fcc phase or $FePt_3$ except FePt fct phase because they contained more Pt than Fe. Also, the amount of soft phase for samples 4# and 5# were about 28.1% and 37.6%, respectively. Sample 2# and sample 3#, which showed almost smooth curve near the magnetization, had almost 0% decrease at the kink. This conclusion was consistent with the XRD results qualitatively.

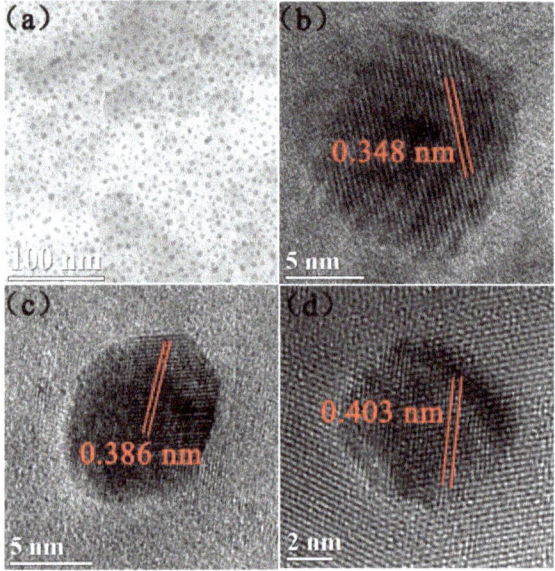

Figure 4. (a) The top-view TEM images of the samples: 3#; the HRTEM images of samples: (b) 1#; (c) 3# and (d) 5#.

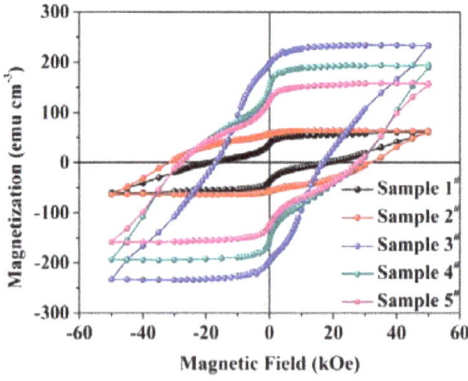

Figure 5. Out-plane field dependence of magnetization recorded at 300K for FePt/MgO nanocomposite films.

Figure 6 shows the squareness and saturation magnetization (M_s) for samples $1^{\#}$ to $5^{\#}$. The saturation magnetization of sample $3^{\#}$ was 234.1 emu/cm^2. The saturation magnetization decreased as the pulsed laser energy density changed. For Fe-rich samples $1^{\#}$ and $2^{\#}$, the saturation magnetizations were 60.7 and 64.6 emu/cm^2, respectively. For samples $4^{\#}$ and $5^{\#}$, which contained more Pt, the saturation magnetizations were 194.8 and 159.1 emu/cm^2, respectively. The variation trend of the saturation magnetization was consistent with the ordering parameter calculated by XRD patterns. The higher the ordering parameter was, the larger was the saturation magnetization. This can be attributed to the soft magnetic phase which reduces the mean atomic magnetic moment. Therefore the samples with lower ordering parameter had less saturation magnetization [29]. It can also be deduced that the FePt/MgO nanocomposite film with almost equal Fe-Pt ratio had larger saturation magnetization than those with unequal Fe-Pt ratio. Moreover the saturation magnetizations of Pt-rich FePt/MgO nanocomposite films were larger than those of Fe-rich samples. This was because to that a small quantity of Pt-rich sample improved the ordering of FePt [11].

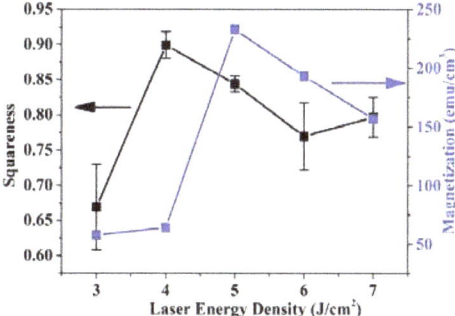

Figure 6. The squareness and saturation magnetization (M_s) of FePt/MgO nanocomposite films.

The squareness ratio (M_r/M_s) of FePt/MgO nanocomposite films is shown in Figure 6. It can be seen that the squareness ratios of samples $2^{\#}$ and $3^{\#}$ were larger than 0.8, and were higher compared to other samples. This demonstrates that the equal Fe-Pt ratio was beneficial with improving the squareness ratio. It is worth noting that sample $2^{\#}$ exhibited large squareness ratio and small saturation magnetization. This was possibly because the Pt content in sample $2^{\#}$ was less than that in sample $3^{\#}$, which made the ordering process incomplete. Adjusting the squareness ratio is helpful for reduction of media noise in magnetic recording media [27].

The magnetic properties of FePt/MgO nanocomposite films indicated that the method used in this study was feasible. By adjusting the ratio of Fe and Pt in the nanocomposite films, the saturation magnetization and coercivity can be tuned at high values. Thus the saturation magnetization values can be controlled without sacrificing much coercivity, which can provide advanced magnets for future applications in high density power and date storage [30–32].

4. Conclusions

In conclusion, PLD was used to fabricate FePt alloy films with different contents of Fe and Pt. The proportion of components was controlled between $Fe_{67}Pt_{33}$ to $Fe_{37}Pt_{63}$ by adjusting the pulsed laser energy density. When the ratio of Fe and Pt was almost 1, the ordering parameter reached 0.9. Furthermore, with the increase in the content of Fe and Pt, the characteristic diffraction peak (001) of FePt shifted to lower and higher angles, respectively. The films with Fe to Pt ratio of almost showed large squareness ratio, while the slightly Pt-rich film had the largest saturation magnetization. Modifying the magnetic properties by tuning the proportion of components via PLD would be helpful for reduction of media noise in magnetic recording media.

Supplementary Materials: The following are available online at http://www.mdpi.com/2079-4991/9/1/53/s1, Figure S1: The TEM image of sample 3#, Table S1: The EDX results of sample 3#.

Author Contributions: Date curation, formal analysis and writing – original draft, J.Y.; Methodology, X.W. (Xuemin Wang); Date curation, T.X.; Methodology, X.Z.; Formal analysis, X.W. (Xinming Wang); Investigation, L.P.; Funding acquisition, Y.Z.; Investigation, J.W.; Data curation, J.C.; Data curation, H.Y.; Funding acquisition and Supervision, W.W.

Funding: This work was financially supported by the National Science Foundation of China (Grant No. 11404302), Laser Fusion Research center funds for young talents and the foundation of the China Academy of Engineering Physics.

Conflicts of Interest: The authors declare no conflict of interest.

References

1. Chen, J.S.; Lim, B.C.; Ding, Y.F.; Chow, G.M. Low-temperature deposition of L1(0) FePt films for ultra-high density magnetic recording. *J. Magn. Magn. Mater.* **2006**, *303*, 309–317. [CrossRef]
2. Chen, J.S.; Hu, J.F.; Lim, B.C.; Ding, Y.F.; Chow, G.M.; Ju, G. Development of L1(0) FePt:C (001) Thin Films With High Coercivity and Small Grain Size for Ultra-High-Density Magnetic Recording Media. *IEEE Trans. Magn.* **2009**, *45*, 839–844. [CrossRef]
3. Giannopoulos, G.; Speliotis, T.; Li, W.F.; Hadjipanayis, G.; Niarchos, D. Structural and magnetic properties of L10/A1, FePt nanocomposites. *J. Magn. Magn. Mater.* **2013**, *325*, 75–81. [CrossRef]
4. Xia, A.; Ren, S.; Lin, J.; Ma, Y.; Xu, C.; Li, J.; Jin, C.; Liu, X. Magnetic properties of sintered $SrFe_{12}O_{19}$-$CoFe_2O_4$ nanocomposites with exchange coupling. *J. Alloys Compd.* **2015**, *653*, 108–116. [CrossRef]
5. Song, F.; Shen, X.; Liu, M.; Xiang, J. Microstructure, magnetic properties and exchange-coupling interactions for one-dimensional hard/soft ferrite nanofibers. *J. Solid State Chem.* **2012**, *185*, 31–36. [CrossRef]
6. Xiong, Z.W.; Cao, L.H. Red-ultraviolet photoluminescence tuning by Ni nanocrystals in epitaxial $SrTiO_3$ matrix. *Appl. Surf. Sci.* **2018**, *445*, 65–70. [CrossRef]
7. Joseyphus, R.J.; Shinoda, K.; Sato, Y.; Tohji, K.; Jeyadevan, B. Composition controlled synthesis of fcc-FePt nanoparticles using a modified polyol process. *J. Mater. Sci.* **2008**, *43*, 2402–2406. [CrossRef]
8. Lu, S. On the tetragonality of martensites in ferrous shape memory alloy Fe_3Pt: A first-principles study. *Acta Mater.* **2016**, *111*, 56–65. [CrossRef]
9. Maat, S.; Kellock, A.J.; Weller, D.; Baglin, J.E.E.; Fullerton, E.E. Ferromagnetism of $FePt_3$ films induced by ion-beam irradiation. *J. Magn. Magn. Mater.* **2003**, *265*, 1–6. [CrossRef]
10. Zhang, Y.; Zhao, L.; Li, S.; Liu, M.; Feng, M.; Li, H. Microstructure and magnetic properties of fcc-FePt/L1(0)-FePt exchange-coupled composite films. *Appl. Phys. A: Mater. Sci. Process.* **2018**, *124*. [CrossRef]
11. Sun, S.H.; Murray, C.B.; Weller, D.; Folks, L.; Moser, A. Monodisperse FePt nanoparticles and ferromagnetic FePt nanocrystal superlattices. *Science* **2000**, *287*, 1989–1992. [CrossRef] [PubMed]
12. Lin, K.W.; Guo, J.Y.; Liu, C.Y.; Ouyang, H.; van Lierop, J.; Phuoc, N.N.; Suzuki, T. Exchange coupling in FePt-$FePt_3$ nanocomposite films. *Phys. Status Solidi A* **2007**, *204*, 3991–3994. [CrossRef]
13. Suber, L.; Imperatori, P.; Bauer, E.M.; Porwal, R.; Peddis, D.; Cannas, C.; Ardu, A.; Mezzi, A.; Kaciulis, S.; Notargiacomo, A.; et al. Tuning hard and soft magnetic FePt nanocomposites. *J. Alloys Compd.* **2016**, *663*, 601–609. [CrossRef]
14. Zeynali, H.; Sebt, S.A.; Arabi, H.; Akbari, H.; Hosseinpour-Mashkani, S.M.; Rao, K.V. Synthesis and Characterization of FePt/NiO Core-Shell Nanoparticles. *J. Inorg. Organomet. Polym. Mater.* **2012**, *22*, 1314–1319. [CrossRef]
15. Nguyen Hoang, N.; Nguyen Thi Thanh, V.; Nguyen Dang, P.; Tran Thi, H.; Nguyen Hoang, H.; Nguyen Hoang, L. Magnetic Properties of FePt Nanoparticles Prepared by Sonoelectrodeposition. *J. Nanomater.* **2012**. [CrossRef]
16. Kong, J.-Z.; Gong, Y.-P.; Li, X.-F.; Li, A.-D.; Zhang, J.-L.; Yan, Q.-Y.; Wu, D. Magnetic properties of FePt nanoparticle assemblies embedded in atomic-layer-deposited Al_2O_3. *J. Mater. Chem.* **2011**, *21*, 5046–5050. [CrossRef]
17. Zheng, B.J.; Lian, J.S.; Zhao, L.; Jiang, Q. Optical and electrical properties of In-doped CdO thin films fabricated by pulse laser deposition. *Appl. Surf. Sci.* **2010**, *256*, 2910–2914. [CrossRef]

18. Gao, Y.; Zhang, X.W.; Yin, Z.G.; Qu, S.; You, J.B.; Chen, N.F. Magnetic Properties of FePt Nanoparticles Prepared by a Micellar Method. *Nanoscale Res. Lett.* **2010**, *5*, 1–6. [CrossRef]
19. Yu, J.; Xiao, T.; Wang, X.; Zhao, Y.; Li, X.; Xu, X.; Xiong, Z.; Wang, X.; Peng, L.; Wang, J.; et al. Splitting of the ultraviolet plasmon resonance from controlling FePt nanoparticles morphology. *Appl. Surf. Sci.* **2018**, *435*, 1–6. [CrossRef]
20. Traub, M.C.; Biteen, J.S.; Michalak, D.J.; Webb, L.J.; Brunschwig, B.S.; Lewis, N.S. Phosphine Functionalization of GaAs(111)A Surfaces. *J. Phys. Chem. C* **2008**, *112*, 18467–18473. [CrossRef]
21. Christodoulides, J.A.; Farber, P.; Daniil, M.; Okumura, H.; Hadjipanayis, G.C.; Skumryev, V.; Simopoulos, A.; Weller, D. Magnetic, structural and microstructural properties of FePt/M (M = C, BN) granular films. *IEEE Trans. Magn.* **2001**, *37*, 1292–1294. [CrossRef]
22. Kim, J.S.; Koo, Y.M. Thickness dependence of (001) texture evolution in FePt thin films on an amorphous substrate. *Thin Solid Films* **2008**, *516*, 1147–1154. [CrossRef]
23. Barmak, K.; Wang, B.C.; Jesanis, A.T.; Berry, D.C.; Rickman, J.M. L1(0) FePt: Ordering, Anisotropy Constant and Their Relation to Film Composition. *IEEE Trans. Magn.* **2013**, *49*, 3284–3291. [CrossRef]
24. Xiao, T.; Wang, X.; Yu, J.; Peng, L.; Zhao, Y.; Xiong, Z.; Shen, C.; Jiang, T.; Yang, Q.; Wang, X.; et al. The microstructure, strain state and optical properties of FePt nano-clusters in MgO matrix. *J. Alloys Compd.* **2018**, *731*, 554–559. [CrossRef]
25. Rellinghaus, B.; Stappert, S.; Acet, M.; Wassermann, E.F. Magnetic properties of FePt nanoparticles. *J. Magn. Magn. Mater.* **2003**, *266*, 142–154. [CrossRef]
26. Tamada, Y.; Morimoto, Y.; Yamamoto, S.; Takano, M.; Nasu, S.; Ono, T. Effects of annealing time on structural and magnetic properties of L1(0)-FePt nanoparticles synthesized by the SiO_2-nanoreactor method. *J. Magn. Magn. Mater.* **2007**, *310*, 2381–2383. [CrossRef]
27. Padmanapan, S. Magnetic properties of FePt based nanocomposite thin films grown on low cost substrates. *Phys. Procedia* **2014**, *54*, 23–29. [CrossRef]
28. Yan, M.L.; Li, X.Z.; Gao, L.; Liou, S.H.; Sellmyer, D.J.; van de Veerdonk, R.J.M.; Wierman, K.W. Fabrication of nonepitaxially grown double-layered FePt: C/FeCoNi thin films for perpendicular recording. *Appl. Phys. Lett.* **2003**, *83*, 3332–3334. [CrossRef]
29. Tsai, J.-L.; Huang, J.-C.; Tai, H.-W.; Tsai, W.-C.; Lin, Y.-C. Magnetic properties and microstructure of FePtB, FePt(B-Ag) granular films. *J. Magn. Magn. Mater.* **2013**, *329*, 6–13. [CrossRef]
30. Yu, Y.; Sun, K.; Tian, Y.; Li, X.Z.; Kramer, M.J.; Sellmyer, D.J.; Shield, J.E.; Sun, S. One-Pot Synthesis of Urchin-like FePd-Fe_3O_4 and Their Conversion into Exchange-Coupled L1(0)-FePd-Fe Nanocomposite Magnets. *Nano Lett.* **2013**, *13*, 4975–4979. [CrossRef] [PubMed]
31. Xiong, Z.; Cao, L. Interparticle spacing dependence of magnetic anisotropy and dipolar interaction of Ni nanocrystals embedded in epitaxial $BaTiO_3$ matrix. *Ceram. Int.* **2018**, *44*, 8155–8160. [CrossRef]
32. Wang, J.; Wang, X.; Yu, J.; Xiao, T.; Peng, L.; Fan, L.; Wang, C.; Shen, Q.; Wu, W. Tailoring the Grain Size of Bi-Layer Graphene by Pulsed Laser Deposition. *Nanomaterials* **2018**, *8*, 885. [CrossRef] [PubMed]

© 2019 by the authors. Licensee MDPI, Basel, Switzerland. This article is an open access article distributed under the terms and conditions of the Creative Commons Attribution (CC BY) license (http://creativecommons.org/licenses/by/4.0/).

Article

InPBi Quantum Dots for Super-Luminescence Diodes

Liyao Zhang [1], Yuxin Song [2,3,*], Qimiao Chen [4], Zhongyunshen Zhu [2] and Shumin Wang [2,3,5,*]

[1] Department of Physics, University of Shanghai for Science and Technology, Shanghai 200093, China; zhangly9@outlook.com
[2] State Key Laboratory of Functional Materials for Informatics, Shanghai Institute of Microsystem and Information Technology, Shanghai 200050, China; phzzys@gmail.com
[3] Key Laboratory of Terahertz Technology, Chinese Academy of Sciences, Shanghai 200050, China
[4] School of Electrical and Electronic Engineering, Nanyang Technological University, Singapore 639798, Singapore; chenqm@ntu.edu.sg
[5] Department of Microtechnology and Nanoscience, Chalmers University of Technology, Göteborg 41296, Sweden
* Correspondence: songyuxin@gmail.com (Y.S.); shumin@mail.sim.ac.cn (S.W.); Tel.: +86-176-1217-8151 (Y.S.); +86-181-0188-6098 (S.W.)

Received: 16 August 2018; Accepted: 6 September 2018; Published: 10 September 2018

Abstract: InPBi thin film has shown ultra-broad room temperature photoluminescence, which is promising for applications in super-luminescent diodes (SLDs) but met problems with low light emission efficiency. In this paper, InPBi quantum dot (QD) is proposed to serve as the active material for future InPBi SLDs. The quantum confinement for carriers and reduced spatial size of QD structure can improve light emission efficiently. We employ finite element method to simulate strain distribution inside QDs and use the result as input for calculating electronic properties. We systematically investigate different transitions involving carriers on the band edges and the deep levels as a function of Bi composition and InPBi QD geometry embedded in InAlAs lattice matched to InP. A flat QD shape with a moderate Bi content of a few percent over 3.2% would provide the optimal performance of SLDs with a bright and wide spectrum at a short center wavelength, promising for future optical coherence tomography applications.

Keywords: InPBi; quantum dot; finite element method; super-luminescent diode; emission spectrum

---

1. Introduction

Dilute bismide is a new member of the group III-V compound semiconductors and has drawn extensive attention for its potential applications in infrared lasers, solar cells and spintronic devices [1–4]. Bismuth (Bi) incorporation can introduce bandgap reduction [5–9], increase the spin orbit splitting energy and reduce the temperature sensitivity of the bandgap [10,11]. As one member of the dilute bismides, InPBi was first theoretically predicted in 1988 [12] and experimentally realized in 2013 [13]. InPBi exhibits broad and strong photoluminescence (PL) at room temperature, ranging between 1.4 and 2.7 µm, making it a potential candidate for fabricating super-luminescent diodes (SLDs) applied in optical coherence tomography (OCT). Deep level transient spectroscopy (DLTS) has confirmed that there are two deep levels in InPBi thin film, a P_{In} antisite donor-like level and a Bi-related acceptor-like level [14]. The multiple PL peaks forming a broad spectrum originate from transitions between carriers in conduction and valence bands and the aforementioned deep levels [14]. Attempts to make SLDs out of InPBi thin film and quantum wells were carried out in our group but revealed weak electroluminescence. One probable reason is that the InPBi grown at low temperature was *n*-type with the electron concentration in the order of 10^{17}–10^{18} cm^{-3} [15]. For the fabricated *p-n* junction, only a small portion of the depletion region lies in the InPBi layer, subsequently, most of the electron-hole recombination happens out of the InPBi region. Another reason is the weak carrier confinement, in particular for electrons. Recent work has shown that Bi atoms distribute inhomogeneously in the InPBi thin films [16]. There are Bi-rich V-shape nanoscale features at the bottom of the InPBi layer close to the InPBi/InP interface and the PL intensity per thickness varies in the InPBi thin film with the maximum value obtained close to the interface where the nanoscale features are observed. This fact indicates that the major contribution to PL in the InPBi thin film is from the bottom part and it is essential to control this part of InPBi epitaxy with Bi-rich nanostructures.

Quantum dot (QD) is one of the most utilized semiconductor nanostructures which confine both electrons and holes and possess discrete energy levels and density of states. Furthermore, the carrier confinement effect can dramatically enhance the radiative recombination efficiency. Due to the outstanding optical properties, semiconductor QDs have been successfully commercialized in telecom lasers [17], visible wavelength LEDs [18] and so forth. In this work, we propose to use InPBi QDs as the active region for InPBi based SLDs. Thanks to the improved carrier confinement and QD size engineering, the InPBi QD device is expected to outperform the similar device made of InPBi thin layer or quantum wells.

III-V QDs are usually grown by the Stranski-Krastanov (SK) mode, triggered by lattice mismatch [19,20]. However, for the $InP_{1-x}Bi_x$/InP system with Bi concentration of a few percent, the lattice mismatch is too small to initiate the SK mode growth. Droplet epitaxy was first realized by Koguchi et al. in the early 1990s [21]. It was introduced to grow III-Vs on II-VIs with almost no lattice mismatch. It usually contains two steps: first, deposit of the group III atoms with the absence of the group V atoms, forming group III metal droplets on the surface and second, exposure of the droplets to the group V atoms and crystallization. Droplet epitaxy provides an effective method to experimentally realize InPBi QDs on materials lattice matched to InP substrate.

In this paper, finite element method (FEM) was employed to simulate the strain distribution in InPBi QDs/InAlAs/InP system and calculate energy levels of electrons and holes. First, the strain distribution in InPBi QDs/InP is simulated. The in-plane strain component is then calculated with different Bi contents and geometries of InPBi QDs. Afterwards, $In_{0.52}Al_{0.48}As$, lattice matched to InP is used to introduce additional potential barrier for InPBi QDs. The strain distribution of the InPBi QDs/InAlAs structure is simulated and utilized as the input for calculations of the ground states of the electrons and holes. Through controlling the size and composition of InPBi QDs, the ground states of the electrons and holes can be engineered. This work provides a feasible way to fabricate SLDs based on InPBi QDs for the potential application in OCT.

2. Methods

Figure 1a shows the schematic of the proposed InPBi QD structure. The $InP_{1-x}Bi_x$ QD is assumed to be in a spherical crown shape with the height varied from 3 to 20 nm and the diameter varied from 20 to 60 nm, according to the typical III-V QDs grown by droplet epitaxy method. The QD is buried in InP or InAlAs for different calculations. The Bi content (x) is varied from 1% to 12%. FEM was employed to simulate the strain distribution in the InPBi QDs/InP system and calculate the energy levels of electrons and holes in the InPBi QDs/InAlAs structure. The lattice constants of InP (a_{InP}) and InBi (a_{InBi}) are 5.87 Å [22] and 6.52 Å [13], respectively, and the lattice constant of $InP_{1-x}Bi_x$ (a_{InPBi}) is $(1-x)a_{InP} + xa_{InBi}$ based on Vegard's law assumption. The lattice mismatch between $InP_{1-x}Bi_x$ and InP is:

$$\text{mis} = \frac{a_{InPBi} - a_{InP}}{a_{InPBi}} = \frac{x}{9+x} \tag{1}$$

The elastic coefficients C11, C12 and C44 of InP and InBi are 1011 GPa, 561 GPa, 456 GPa [22] and 60.31 GPa, 32.52 GPa [12], 26.1 GPa [23], respectively. The elastic coefficient of $InP_{1-x}Bi_x$ is assumed as:

$$C_{ij}(InP_{1-x}Bi_x) = (1-x)C_{ij}(InP) + xC_{ij}(InBi) \tag{2}$$

The elastic coefficients C11, C12 and C44 of $In_{0.52}Al_{0.48}As$ is assumed to be the linear combination of that of InAs and AlAs, which is deduced to be 1033 GPa, 492 GPa and 466 GPa [22], respectively.

The conduction band hydrostatic deformation potential a_c and valence band hydrostatic deformation potential a_v of InP and In$_{0.52}$Al$_{0.48}$As are −6.0 eV and −0.6 eV [22] and −6.7 eV and −0.8 eV [24], respectively. Due to the lack of available data, the a_c and a_v of InPBi are assumed to be equal to that of InP with the consideration that the Bi content is small. Then the ground states of electrons, heavy holes and light holes are calculated by Schrödinger equation with the simulated strain distribution as an input.

$$\left[\frac{-\hbar^2}{2m*}\nabla^2 + V(r)\right]\Psi(r) = E\Psi(r) \tag{3}$$

$$V(r) = V_e(r) + V_s(r) \tag{4}$$

$$V_s(r) = a_{c,v}(\varepsilon_{xx} + \varepsilon_{yy} + \varepsilon_{zz}) \tag{5}$$

where $\hbar$ is the reduced Planck constant and $m*$ is the effective mass of the carriers. For In$_{0.52}$Al$_{0.48}$As, the effective mass of electron (m_e^*) heavy hole (m_{hh}^*) and light hole (m_{lh}^*) is $0.069m_0$, $0.4m_0$ and $0.103m_0$ [22], respectively, where m_0 is the mass of electron. For InPBi, the effective mass of InP is used instead, that is $0.0795m_0$, $0.6m_0$ and $0.089m_0$ for electron, heavy hole and light hole [22], respectively. $V(r)$ is the potential, including the potential $V_e(r)$ caused by intrinsic band offset and the potential $V_s(r)$ introduced by strain. The energy of the InP conduction band minima is set to be 0. Without consideration of strain, the potential $V_e(r)$ of the electrons and holes of In$_{0.52}$Al$_{0.48}$As are 0.34 eV and −1.17 eV, respectively. Because the bandgap reduction rate of conduction band and valence band of InPBi is −27 meV/%Bi and 79 meV/%Bi [9], respectively, the potential $V_e(r)$ of the electrons and holes of InP$_{1-x}$Bi$_x$ is −0.027x eV and (−1.42 + 0.079x) eV, respectively. The parameters used for the calculations are summarized in Table 1.

Table 1. Summary of the parameters used for the calculations.

Parameters	InP [22]	InBi	In$_{0.52}$Al$_{0.48}$As
C11 (GPa)	1011	60.31 [12]	1033 [22]
C12 (GPa)	561	32.52 [12]	492 [22]
C44 (GPa)	456	26.1 [23]	466 [22]
a_c (eV)	−6		−6.7 [24]
a_v (eV)	−0.6		−0.8 [24]
m_e^*	$0.0795m_0$		$0.069m_0$ [22]
m_{hh}^*	$0.6m_0$		$0.4m_0$ [22]
m_{lh}^*	$0.089m_0$		$0.103m_0$ [22]

3. Results and Discussions

3.1. Strain Analysis

At first, strain analysis is carried out for the most simplified prototype model, an InPBi QD buried in InP, as shown in Figure 1a. The simulated InPBi QD has a Bi content of 6%, a diameter of 40 nm and a height of 6 nm. Figure 1b–e shows the distribution of various strain components, ε_{xx}(b) and ε_{zz}(c) in the yz plane cross the center of the QD and ε_{xy}(d) and ε_{xz}(e) in the xy plane cross the bottom, respectively. The deformation is exaggerated by 100 times. Because the lattice constant of InPBi is larger than that of InP, the InPBi QD tends to expand, as seen in Figure 1b–e. The ε_{xx} in the yz plane is negative, indicating compressive in-plane strain in the InPBi QD. The ε_{xx} distributes quite uniformly within the QD with an average value of -5.82×10^{-3}. The strain in InP around the InPBi QD is tensile and gradually decreases to 0 when it is about 10 nm away from the QD. The ε_{zz} in the yz plane is positive in the InPBi QD, indicating tensile strain in the z direction. ε_{zz} in the InP above and below the InPBi QD is negative but the strain around the edge of the QD is positive. The shear strain components ε_{xy} (d) and ε_{xz} (e) are asymmetric in the xy plane. The in-plane strain ε_{xx}, the vertical strain ε_{zz} and the shear strain ε_{xz} are larger than the shear strain ε_{xy} for about one order of magnitude.

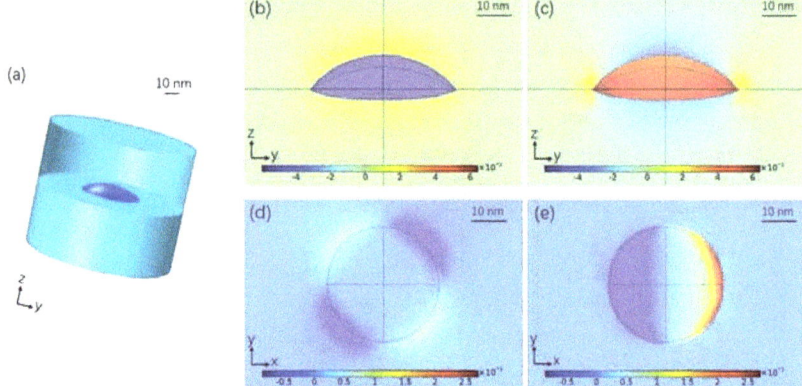

Figure 1. (a) Three-dimensional schematic of the proposed InPBi QD structure. The strain distribution of (b) ε_{xx} and (c) ε_{zz} in the yz plane; (d) ε_{xy} and (e) ε_{xz} in the xy plane for the InPBi QDs in InP with Bi content of 6%, diameter of 40 nm and height of 6 nm. The deformation is exaggerated by 100 times.

Afterwards, the influence of Bi content and geometric parameters on strain distribution is systemically investigated. Figure 2 shows the variation of the in-plane strain ε_{xx} versus the Bi content (x) (a) and the aspect ratio (D/H) (b) of the InPBi QD. The ε_{xx} shown in Figure 2a is averaged over the whole InPBi QD with a fixed diameter of 40 nm and height of 6 nm, while in Figure 2b the ε_{xx} is calculated with Bi content fixed at 6%. The D and H designate the diameter and height of the InPBi QDs, respectively. It can be found in Figure 2a that the average in-plane strain ε_{xx} is almost linear to the Bi content with a slope of -0.097. Thus

$$\varepsilon_{xx} = -0.097x \tag{6}$$

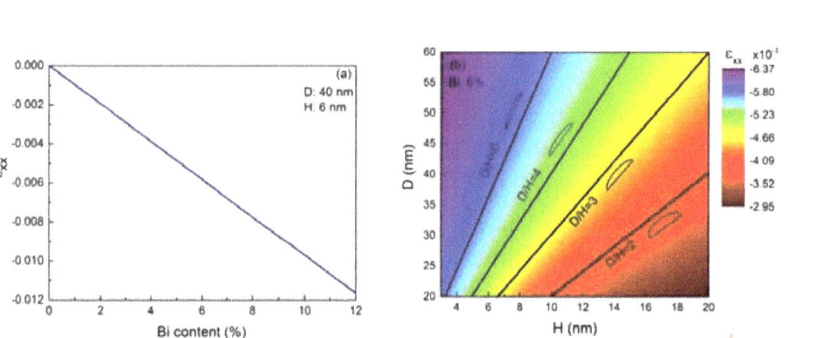

Figure 2. (a) The simulated average in-plane strain ε_{xx} in the InPBi QDs versus the Bi content with fixed diameter and height of the QDs of 40 nm and 6 nm, respectively; (b) contour map of the average ε_{xx} versus the diameter and height of the InPBi QDs with the Bi content of 6%. The black lines represent the aspect ratio of 2, 3, 4 and 6, respectively, with the diagrammatic sketch of the shape of the QD next to each line.

The average in-plane strain ε_{xx} is compressive and its amplitude increases with the Bi content. Figure 2b is a contour map of the average ε_{xx} versus the diameter and height of the InPBi QDs together with a few representative lines on which the aspect ratio is the same. The average in-plane strain ε_{xx} is found proportional to the aspect ratio of the InPBi QDs. The diameter and height of the InPBi QDs vary from 20 nm to 60 nm and 3 nm to 20 nm, respectively, thus, the aspect ratio varies from 1 to 20. The average in-plane strain ε_{xx} then varies from -2.95×10^{-3} to -6.37×10^{-3}. For InPBi

QDs with different diameters and heights, as long as the aspect ratio is the same, the in-plane strain ε_{xx} is also the same. The higher D/H ratio, or in another word, the flatter the QD is, the larger the average ε_{xx}.

3.2. Band Structure

In order to enhance the carrier confinement, $In_{0.52}Al_{0.48}As$ lattice matched to InP is chosen to replace the InP around the InPBi QDs to provide potential barriers for both conduction band and valence band. Thus, the hetero-structure becomes InAlAs/InPBi QDs/InAlAs.

Figure 3a shows the calculated band structure of an InAlAs/InPBi QD/InAlAs hetero-structure, with the diameter, the height and the Bi content of the InPBi QD of 40 nm, 6 nm and 6%, respectively. The bottom of the conduction band and the top of the valence band of $InP_{0.94}Bi_{0.06}$ QD lies at -0.12 eV and -0.94 eV, respectively. The blue, red and green lines represent the ground state of the electrons (-0.017 eV), the heavy holes (-1.00 eV) and the light holes (-1.01 eV), respectively. The magenta and purple dashed lines represent the P_{In} antisite level and the Bi-cluster-related level, respectively. According to our former DLTS measurements, there are two deep levels in InPBi thin films. The P_{In} antisite level lies at 0.31 eV below the conduction band of InPBi and the Bi-cluster-related level lies at 0.11 eV above the valence band of InPBi for the InPBi thin film with Bi content of 2.49% [14]. Three radiative recombination processes involving the deep levels were identified from PL measurements. The first is the recombination between the electrons in the conduction band and the holes at the Bi-cluster-related level, marked as HE, the second is between the electrons at the P_{In} antisite level and the holes in the valence band, marked as ME and the third is the recombination between the two deep levels, marked as LE. These three carrier recombination processes together result in very broad PL spectra of InPBi thin films. In the InPBi QD case, due to the quantum confinement, the energy levels of electrons and holes are split into discrete energy levels. Consequently, the recombination with the deep levels will involve the energy levels instead of the band edges of InPBi. The arrows labeled HE, ME and LE in Figure 3a indicate the expected recombination processes in the InPBi QDs. Since the ground state of the heavy holes has lower energy than that of the light holes, the ME process is expected to be between the P_{In} antisite level and the ground state of the heavy holes. The ΔE_{HE} and the ΔE_{ME} are the energy difference between the electron ground state and the conduction band edge of $InP_{0.94}Bi_{0.06}$ and between the valence band edge and the heavy hole ground state, respectively.

The variation of the ground state of the electrons, the heavy holes and the light holes and the energy difference ΔE_{HE} and ΔE_{ME} with the Bi content and the height of the InPBi QDs are further investigated, as shown in Figure 3b–e. In Figure 3b,c, the diameter and the height of the InPBi QDs are fixed at 40 nm and 6 nm, respectively, while In Figure 3d,e, the diameter and the Bi content of the InPBi QDs are fixed at 40 nm and 6%, respectively. The blue, red, green, magenta and purple curves represent the ground state of electrons, heavy holes and light holes and the energy difference ΔE_{HE} and ΔE_{ME}, respectively. The ground state energy of the electrons is found monotonically decrease with the Bi content varying from 1% to 12%. The absolute value of the ground state energies of the heavy holes and the light holes also monotonically decrease with the Bi content varying from 3.2% to 12%. The difference between the ground state energy of the heavy holes and the light holes increases with the Bi content. When the Bi content is below 3.2%, the valence band edge of InPBi is lower than that of $In_{0.52}Al_{0.48}As$ and thus there is no potential well for the holes in InPBi, neither the ground states. The dependence of the ground state energies on Bi content is dominant by the bandgap reduction of InPBi. The energy difference ΔE_{HE} and ΔE_{ME} are found also monotonically increase with the Bi content. The slope of the ΔE_{HE}–Bi content curve is almost uniform when the Bi content varies from 1% to 12%. However, the ΔE_{ME} changes merely when the Bi content is between 4% and 12% and sharply drops when the Bi content decreases below 4%. The dependence of ΔE_{HE} and ΔE_{ME} on Bi content is mainly caused by the fact that when the Bi content increases, both the conduction and valence band offset increase and the ground states of the carriers are elevated relative to the band edges. The valence band offset is more influenced by the Bi content, even changing from type-I to type-II band alignment at about 3.2%.

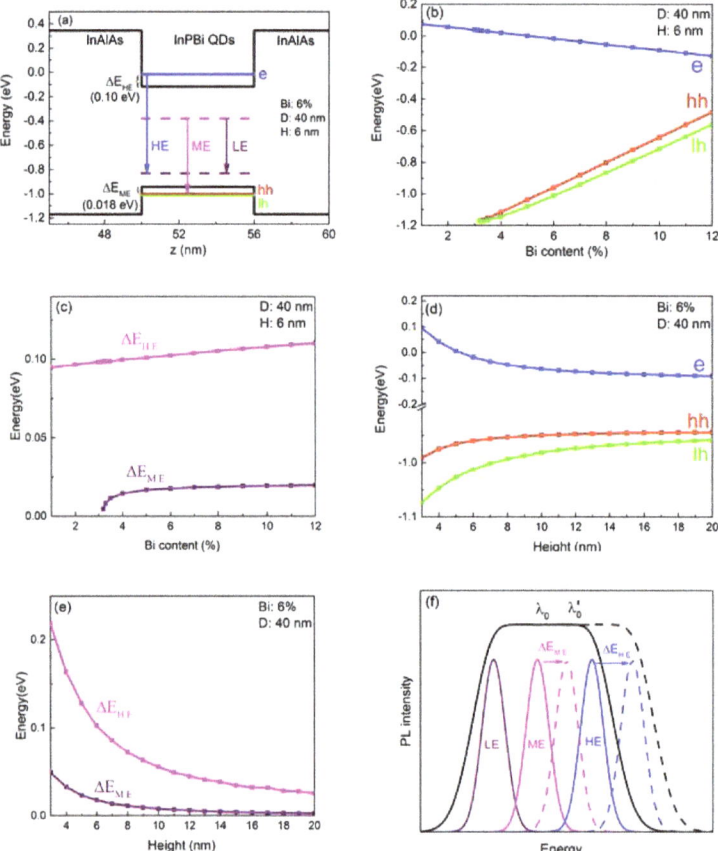

Figure 3. (a) the band alignment and carrier recombination processes of an InAlAs/InPBi QD structure plotted along the z axis across the center of the InPBi QD with the Bi content, the diameter and the height of 6%, 40 nm and 6 nm, respectively. The zero in Energy is set as the bottom of the conduction band of InP. The blue, red and green line are the ground state of the electrons, heavy holes and light holes, respectively. ΔE_{HE} and ΔE_{ME} are the energy difference between the electron ground state and the conduction band edge of $InP_{0.94}Bi_{0.06}$ and between the valence band edge and the heavy hole ground state, respectively; (b,c) show the dependence of the ground state energy of the electrons (blue), the heavy holes (red) and the light holes (green) (b) and the energy difference ΔE_{HE} (magenta) and ΔE_{ME} (purple) (c) on the Bi content of the InPBi QD with fixed diameter and height of 40 nm and 6 nm, respectively; (d,e) show the dependence of the ground state energy of the electrons (blue), the heavy holes (red) and the light holes (green) (d) and the energy difference ΔE_{HE} (magenta) and ΔE_{ME} (purple) (e) on the height of the InPBi QDs with fixed diameter and Bi content of 40 nm and 6%, respectively. The markers indicate the simulated data points; (f) diagrammatic sketch of the PL spectrum broadening with the energy increase of the ME and HE transitions. The solid lines represent the LE (purple), ME (magenta), HE (blue) and PL (black) of the InPBi thin films. The dash lines are the ME transition, ME transition and PL (black) of the InAlAs/InPBi QD structure.

Next, the influence of the geometric shape of the QDs on the ground states of the carriers is investigated. The ground state energy levels of the electrons, the heavy holes and the light holes are found to decrease with increasing the height of the InPBi QDs from 3 nm to 20 nm, as shown in Figure 3d. The ground state energy of the electrons remarkably drops when the height of the InPBi

QDs increases from 3 nm to 8 nm, resulted from strong quantum confinement when the potential well is thin and then slowly decreases as the height increases from 8 nm to 12 nm. The absolute value of the ground state energy of the heavy holes and the light holes behaves similarly as the electrons. The difference in slopes originates from the difference in effective masses. The energy differences ΔE_{HE} and ΔE_{ME} in (e) have the same trend as the electron and the heavy hole ground states in (d) since the band edges are fixed in this group of simulations.

Figure 3f is the diagrammatic sketch of the PL spectra of InPBi thin films and QDs. The solid curves represent PL contributions of the LE (purple), ME (magenta), HE (blue) and the overall PL (black) of InPBi thin films, while the dashed curves represent the case of the InAlAs/InPBi QD structure, respectively. Compared to InPBi thin films, the energies of the emitted photons from InPBi QDs are increased with the value of ΔE_{HE} and ΔE_{ME} for the HE and ME transitions and the center energy of the spectra will also be increased. The high spatial resolution of OCT requires a spectrum with large linewidth and short center wavelength. Based on the above results, InPBi QDs with a low height can increase the ΔE_{HE} and ΔE_{ME} and thus lead to a wide linewidth as well as a short center wavelength of the emission spectrum, subsequently improving the spatial resolution of OCT.

The ultimate aim of the proposal using InPBi QDs is to produce high performance SLDs with a bright, flat and broad spectrum. The brightness requires a high Bi content to provide large quantum confinement for the carriers. Use of a high Bi content will decrease the bandgap and subsequently decrease the transition energy of HE and ME, leading to a reduced linewidth of the emission spectrum. This can be compensated by controlling the shape of the QDs. Finally, a moderate Bi content of a few percent over 3.2% and a flat QD shape would provide the optimal performance.

Furthermore, unlike the InAs QDs on GaAs platform with large lattice mismatch, the InPBi QDs on InP lattice has limited strain and can thus be stacked for many periods without the risk of strain relaxation. The stacked structure can not only increase the overall intensity but also further engineer the shape and linewidth of the emission spectrum by manipulating the shape and Bi content of the InPBi QDs in each period.

4. Conclusions

In this paper, we propose to use InPBi QD as the active region for high performance SLDs. FEM was employed to calculate the strain distribution of the InP/InPBi QDs/InP structure. The in-plane strain components are found larger than the shear strain components in the InPBi QD. The average in-plane strain ε_{xx} is linear to the Bi content and proportional to the aspect ratio of the InPBi QDs. $In_{0.52}Al_{0.53}As$ lattice matched to InP is chosen to form potential barriers for the carriers in InPBi QDs. The band alignment and the ground state of electrons, heavy holes and light holes are calculated with different Bi contents and heights of the InPBi QDs. High Bi content can reduce the bandgap and deepen the band offset leading to improved quantum confinement and optical property. Low height can increase both the ΔE_{HE} and ΔE_{ME}, especially the ΔE_{HE} and consequently increase the linewidth of the emission spectrum. A moderate Bi content of a few percent over 3.2% and a flat QD shape would provide the optimal performance of SLDs with high light emission efficiency, wide spectrum and shortened center wavelength for future OCT applications.

Author Contributions: L.Z. was responsible for the calculations, analysis of data and preparation of the manuscript. Q.C. and Z.Z. helped with the programming. Y.S. supervised the entire study and revised the manuscript. S.W. participated in the manuscript revision.

Funding: The authors are grateful for the financial support provided by the Open Project Program of the State Key Laboratory of Functional Materials for Informatics (Grant No. SKL-2018-11).

Conflicts of Interest: The authors declare no conflict of interest.

References

1. Sweeney, S.; Jin, S. Bismide-nitride alloys: Promising for efficient light emitting devices in the near-and mid-infrared. *J. Appl. Phys.* **2013**, *113*, 043110. [CrossRef]
2. Marko, I.; Batool, Z.; Hild, K.; Jin, S.R.; Hossain, N.; Hosea, T.J.C.; Petropoulos, J.P.; Zhong, Y.; Dongmo, P.B.; Zide, J.M.O.; et al. Temperature and Bi-concentration dependence of the bandgap and spin-orbit splitting in InGaBiAs/InP semiconductors for mid-infrared applications. *Appl. Phys. Lett.* **2012**, *101*, 221108. [CrossRef]
3. Tiedje, T.; Young, E.C.; Mascarenhas, A. Growth and properties of the dilute bismide semiconductor GaAs1-xBix a complementary alloy to the dilute nitrides. *Int. J. Nanotechnol.* **2008**, *5*, 963–983. [CrossRef]
4. Dimroth, F. High-efficiency solar cells from III-V compound semiconductors. *Phys. Status solidi C* **2006**, *3*, 373–379. [CrossRef]
5. Rajpalke, M.K.; Linhart, W.M.; Birkett, M.; Yu, K.M.; Alaria, J.; Kopaczek, J.; Kudrawiec, R.; Jones, T.S.; Ashwin, M.J.; Veal, T.D. High Bi content GaSbBi alloys. *J. Appl. Phys.* **2014**, *116*, 043511. [CrossRef]
6. Francoeur, S.; Seong, M.J.; Mascarenhas, A.; Tixier, S.; Adamcyk, M.; Tiedje, T. Band gap of GaAs$_{1-x}$Bi$_x$, $0 < x < 3.6\%$. *Appl. Phys. Lett.* **2003**, *82*, 3874–3876.
7. Ma, K.Y.; Fang, Z.M.; Jaw, D.H.; Cohen, R.M.; Stringfellow, G.B.; Kosar, W.P.; Brown, D.W. Organometallic vapor phase epitaxial growth and characterization of InAsBi and InAsSbBi. *Appl. Phys. Lett.* **1989**, *55*, 2420–2422. [CrossRef]
8. Jean-Louis, A.; Hamon, C. Propriétés des alliages InSb$_{1-x}$Bi$_x$ I. Mesures électriques. *Phys. Status Solidi B* **1969**, *34*, 329–340. [CrossRef]
9. Kopaczek, J.; Kudrawiec, R.; Polak, M.P.; Scharoch, P.; Birkett, M.; Veal, T.D.; Wang, K.; Gu, Y.; Gong, Q.; Wang, S. Contactless electroreflectance and theoretical studies of band gap and spin-orbit splitting in InP1−xBix dilute bismide with x ≤ 0.034. *Appl. Phys. Lett.* **2014**, *105*, L1283. [CrossRef]
10. Alberi, K.; Wu, J.; Walukiewicz, W.; Yu, K.M.; Dubon, O.D.; Watkins, S.P.; Wang, C.X.; Liu, X.; Cho, Y.-J.; Furdyna, J. Valence-band anticrossing in mismatched III-V semiconductor alloys. *Phys. Rev. B* **2007**, *75*, 045203. [CrossRef]
11. Fluegel, B.; Francoeur, S.; Mascarenhas, A.; Tixier, S.; Young, E.C.; Tiedje, T. Giant spin-orbit bowing in GaAs$_{1-x}$Bi$_x$. *Phys. Rev. Lett.* **2006**, *97*, 067205. [CrossRef] [PubMed]
12. Berding, M.A.; Sher, A.; Chen, A.B.; Miller, W.E. Structural properties of bismuth-bearing semiconductor alloys. *J. Appl. Phys.* **1988**, *63*, 107–115. [CrossRef]
13. Wang, K.; Gu, Y.; Zhou, H.F.; Zhang, L.Y.; Kang, C.Z.; Wu, M.J.; Pan, W.W.; Lu, P.F.; Gong, Q.; Wang, S.M. InPBi Single Crystals Grown by Molecular Beam Epitaxy. *Sci. Rep.* **2014**, *4*, 5449. [CrossRef] [PubMed]
14. Wu, X.; Chen, X.; Pan, W.; Wang, P.; Zhang, L.; Li, Y.; Wang, H.; Wang, K.; Shao, J.; Wang, S. Anomalous photoluminescence in InP$_{1-x}$Bi$_x$. *Sci. Rep.* **2016**, *6*, 27867. [CrossRef] [PubMed]
15. Pan, W.; Wang, P.; Wu, X.; Wang, K.; Cui, J.; Yue, L.; Zhang, L.; Gong, Q.; Wang, S. Growth and material properties of InPBi thin films using gas source molecular beam epitaxy. *J. Alloys Compd.* **2016**, *656*, 777–783. [CrossRef]
16. Zhang, L.; Wu, M.; Chen, X.; Wu, X.; Spiecker, E.; Song, Y.; Pan, W.; Li, Y.; Yue, L.; Shao, J.; et al. Nanoscale distribution of Bi atoms in InP$_{1-x}$Bi$_x$. *Sci. Rep.* **2017**, *7*, 12278. [CrossRef] [PubMed]
17. Liu, H.; Wang, T.; Jiang, Q.; Hogg, R.; Tutu, F.; Pozzi, F.; Seeds, A. Long-wavelength InAs/GaAs quantum-dot laser diode monolithically grown on Ge substrate. *Nat. Photonics* **2011**, *5*, 416. [CrossRef]
18. Qian, L.; Zheng, Y.; Xue, J.; Holloway, P.H. Stable and efficient quantum-dot light-emitting diodes based on solution-processed multilayer structures. *Nat. Photonics* **2011**, *5*, 543. [CrossRef]
19. Lent, C.S.; Tougaw, P.D.; Porod, W.; Bernstein, G.H. Quantum cellular automata. *Nanotechnology* **1993**, *4*, 49. [CrossRef]
20. Wu, J.; Wang, Z.M. *Quantum Dot Molecules*; Springer: New York, NY, USA, 2014.
21. Chikyow, T.; Koguchi, N. MBE growth method for pyramid-shaped GaAs micro crystals on ZnSe (001) surface using Ga droplets. *Jpn. J. Appl. Phys.* **1990**, *29*, L2093. [CrossRef]
22. Vurgaftman, I.J.; Meyer, J.Á.; Ram-Mohan, L. Band parameters for III–V compound semiconductors and their alloys. *J. Appl. Phys.* **2001**, *89*, 5815–5875. [CrossRef]

23. Shalindar, A.J.; Webster, P.T.; Wilkens, B.J.; Alford, T.L.; Johnson, S.R. Measurement of InAsBi mole fraction and InBi lattice constant using Rutherford backscattering spectrometry and X-ray diffraction. *J. Appl. Phys.* **2016**, *120*, 145704. [CrossRef]
24. Yeh, C.N.; McNeil, L.E.; Nahory, R.E.; Bhat, R. Measurement of the In0.52Al0.48As valence-band hydrostatic deformation potential and the hydrostatic-pressure dependence of the In0.52Al0.48As/InP valence-band offset. *Phys. Rev. B* **1995**, *52*, 14682. [CrossRef]

© 2018 by the authors. Licensee MDPI, Basel, Switzerland. This article is an open access article distributed under the terms and conditions of the Creative Commons Attribution (CC BY) license (http://creativecommons.org/licenses/by/4.0/).

Article

Hierarchical Structure and Catalytic Activity of Flower-Like CeO$_2$ Spheres Prepared Via a Hydrothermal Method

Genli Shen, Mi Liu, Zhen Wang * and Qi Wang *

CAS Key Laboratory of Standardization and Measurement for Nanotechnology, National Center for Nanoscience and Technology, Beijing 100190, China; shengl@nanoctr.cn (G.S.); liumi@nanoctr.cn (M.L.)
* Correspondence: wangzh@nanoctr.cn (Z.W.); wangq@nanoctr.cn (Q.W.);
  Tel.: +86-134-6671-8278 (Z.W); +86-186-0129-2518 (Q.W.)

Received: 24 August 2018; Accepted: 25 September 2018; Published: 29 September 2018

Abstract: Hierarchical CeO$_2$ particles were synthesized by a hydrothermal method based on the reaction between CeCl$_3 \cdot$7H$_2$O and PVP at 270 °C. The flower-like CeO$_2$ with an average diameter of about 1 µm is composed of compact nanosheets with thicknesses of about 15 nm and have a surface area of 36.8 m^2/g, a large pore volume of 0.109 cm^3/g, and a narrow pore size distribution (14.9 nm in diameter). The possible formation mechanism of the hierarchical CeO$_2$ nanoparticles has been illustrated. The 3D hierarchical structured CeO$_2$ exhibited a higher catalytic activity toward CO oxidation compared with commercial CeO$_2$.

Keywords: ceria; catalytic activity; hierarchical structure

1. Introduction

CeO$_2$ is playing important roles in various fields such as promoters for three-way catalysts [1], fuel cells [2], hydrogen storage materials [3], and oxygen sensors [4]. Although the utilization of ceria is based on its intrinsic chemical properties, the structures and morphologies of CeO$_2$ also have a significant influence on its properties and applications [5,6].

So far, CeO$_2$ with different sizes and morphologies such as nanorods [7], nanospheres [8], nanotubes [9], and nanocubes [10] have been synthesized in the last decade. It was proved that CeO$_2$ nanoparticles with different sizes and morphologies have better properties than general CeO$_2$ does. CeO$_2$ nanoparticles afford more active sites because of their high specific surface areas and novel structures [11].

Preparation of CeO$_2$ with different structures and morphologies provides the basic groundwork for its advanced applications. Hierarchical structured CeO$_2$ with unique properties and novel functionalities has attracted the attention of many researchers in recent years.

Zhong et al. synthesized a three-dimensional (3D) flower-like CeO$_2$ micro/nanocomposite structure using cerium chloride as a reactant by a simple and economical route based on an ethylene glycol-mediated process [12]. Li et al. synthesized mesoporous Ce(OH)CO$_3$ microspheres with flower-like 3D hierarchical structures via different hydrothermal systems, including glucose/acylic acid, fructose/acrylic acid, glucose/propanoic acid, and glucose/n-butylamine systems. Calcination of the Ce(OH)CO$_3$ microspheres yielded mesoporous CeO$_2$ microspheres with the same flower-like morphology as that of Ce(OH)CO$_3$ microspheres [13]. Ouyang et al. reported a facile electrochemical method to prepare hierarchical porous CeO$_2$ nanospheres and applied them as highly efficient absorbents to remove organic dyes [14]. However, 3D hierarchical structured CeO$_2$ is commonly synthesized with relatively miscellaneous process, which limited the extensive usage of the prepared ceria. In this paper, we report a facile one-pot hydrothermal route to synthesize 3D hierarchical structured CeO_2. The present hydrothermal route is low cost and can be easily scaled-up. The fabricated 3D hierarchical structured CeO_2 could be used as a catalyst for CO oxidation and a support for noble metal catalysts.

2. Materials and Methods

2.1. Preparation of Hierarchical Structured CeO_2

Cerium (III) chloride heptahydrate ($CeCl_3 \cdot 7H_2O$), polyvinyl pyrrolidone (PVP), and ethanol were purchased from Beijing Yili Chemical Reagent Co. Ltd. (Beijing, China). All materials were used without any further purification. In a typical synthetic procedure of the hierarchical structured CeO_2, 0.5 mmol $CeCl_3 \cdot 7H_2O$ was dissolved in 30 mL deionized water, and then 1 mmol PVP and 20 mL deionized water were added to the solution. After 15 min of magnetic stirring, the homogenous solution was transferred into the Teflon vessel of a hydrothermal bomb, which was then placed in an oven and maintained at 270 °C for 24 h. Then, the solution was cooled to room temperature, and the products were separated by centrifugation and washed with absolute ethanol and distilled water.

2.2. Characterization Techniques

The crystal phases of the products were characterized by X-ray diffraction (XRD) using Philips X'pert PRO analyzer (Philips, Amsterdam, The Netherlands) equipped with a Cu K_α radiation source (λ = 0.154187 nm) and operated at an X-ray tube (Philips, Amsterdam, The Netherlands) voltage and current of 40 KV and 30 mA, respectively. The morphology of the products was examined by scanning electron microscopy (SEM) using a JEOL JSM 67OOF system (JEOL, Tokyo, Japan) and transmission electron microscopy (TEM) using a JEM-2100 system (JEOL, Tokyo, Japan) operated at 200 kV. Surface composition was determined by X-ray photoelectron spectroscopy (XPS) using an ESCALab220i-XL electron spectrometer (VG Scientific, Waltham, MA, USA) with monochromatic Al K_α radiation. Nitrogen adsorption-desorption isotherms were analyzed using an automatic adsorption system (Autosorb-1, Quantachrome, Boynton Beach, FL, USA) at the temperature of liquid nitrogen.

3. Results

3.1. 3D Hierarchical Structured CeO_2 Prepared via Hydrothermal Method

The powder XRD pattern of the as-prepared sample is shown in Figure 1. As can be seen, the as-prepared sample can be indexed to the cubic phase of CeO_2 (JCPDS No. 34-0394). The average crystallite size calculated by the Scherrer equation is 26.8 nm.

The SEM images of the as-synthesized CeO_2 particles are shown in Figure 2. It can be seen from Figure 2a that the as-synthesized CeO_2 microspheres have diameters of about 1 µm. These CeO_2 microspheres consist of many nanosheets with thicknesses of about 15 nm. The mesopores with about 20-nm pore sizes are spread over the nanosheets. The lattice fringes in the high-resolution TEM (HRTEM) image (Figure 2c) show a spacing of 0.31 nm, corresponding to the (1 1 1) plane of cubic CeO_2. The selected area electron diffraction (SAED) pattern (Figure 2d) indicates that the microspheres are composed of low-crystalline CeO_2 nanocrystals.

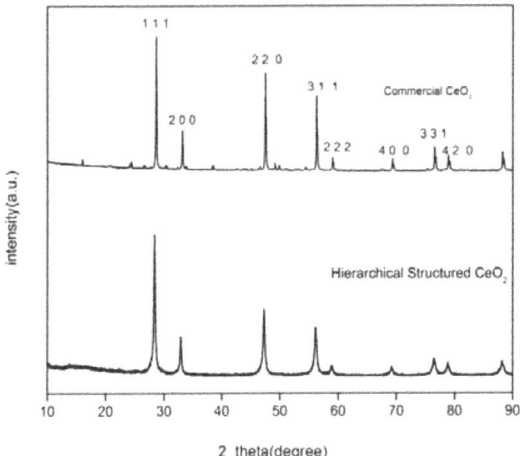

Figure 1. XRD pattern of as-prepared 3D hierarchical structured CeO_2.

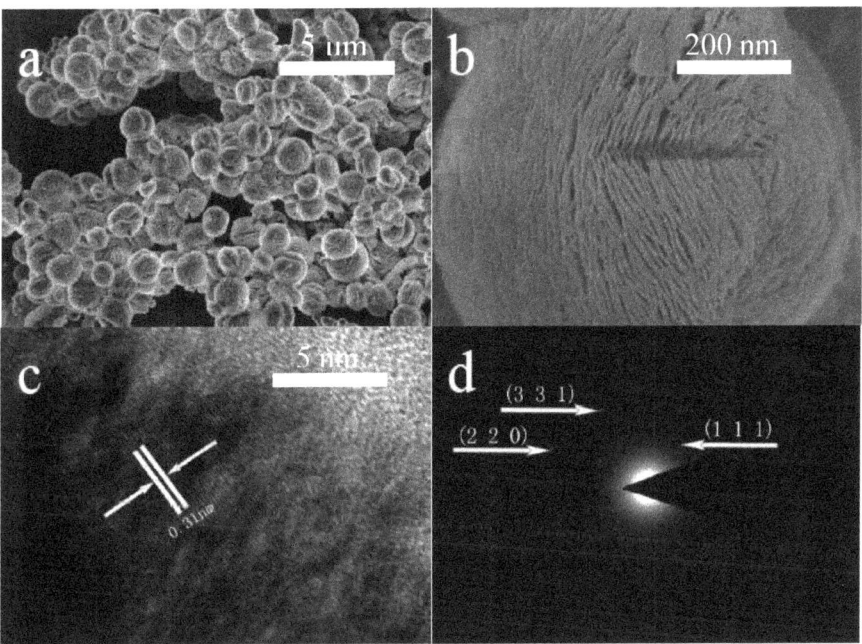

Figure 2. (a,b) SEM images, (c) HRTEM image, and (d) SAED pattern of the 3D hierarchical structured CeO_2.

The nitrogen adsorption and desorption isotherms of the as-prepared samples and the corresponding pore size distribution curve calculated by the Barret-Joyner-Halenda (BJH) method are shown in Figure 3. The nitrogen adsorption and desorption isotherms exhibit a slim hysteresis loop at a relative pressure of >0.2, which is the type-II curve. The calculated Brunauer-Emmett-Teller (BET) surface area of the as-synthesized CeO_2 is about 36.8 m^2g^{-1}. The average pore size calculated by the BJH method is 14.9 nm.

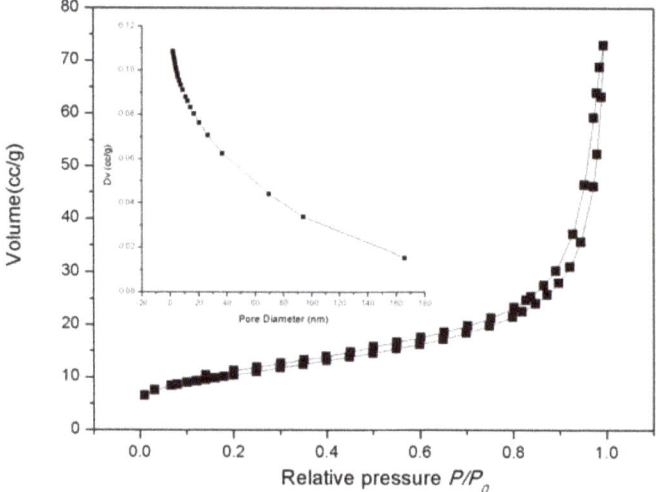

Figure 3. Nitrogen adsorption-desorption isotherm of 3D hierarchical structured CeO_2. The inset shows the pore size distribution curve obtained from the desorption data.

3.2. Effects of Synthesis Conditions on the Formation of 3D Hierarchical Structured CeO_2 and the Possible Formation Mechanism

To investigate the evolution of flower-like CeO_2 particles, the samples obtained after different reaction times were characterized by SEM (Figure 4). The reaction temperature and the dosages of $CeCl_3 \cdot 7H_2O$ and PVP were kept constant (270 °C, 0.01 M, and 0.02 M, respectively). As we can see in Figure 4a, spherical particles were obtained in the early stage. After 12 h of hydrothermal treatment, the sample (Figure 4b) evolved into spheres on which many scrappy grains grew. We speculate that PVP at the surface of the spheres decomposed gradually at such a high temperature and pressure, and simultaneously, tiny nanoparticles on the surface of the spheres began to grow into nanosheets. As seen in Figure 4c, all spheres have transformed into flower-like CeO_2 particles. Based on these observations, the possible formation mechanism of the 3D hierarchical structured CeO_2 can be speculated. The schematic mechanism for the 3D hierarchical structured CeO_2 obtained during different hydrothermal stages is illustrated in Figure 5. At an early stage, Ce^{3+} ions were oxidized gradually by O_2 present in the aqueous solution to form small CeO_2 nanocrystals. Then, the small CeO_2 nanocrystals interacted with PVP and self-assembled as building blocks into spherical particles. As the temperature of the hydrothermal system increased, the PVP at the surface of the spherical particles began to decompose and small nanoparticles began to grow into nanosheets via Ostwald ripening. Due to Ostwald ripening, more were nanosheets formed, and after 24 h of hydrothermal treatment, the PVP completely decomposed and 3D hierarchical structured CeO_2 particles were formed.

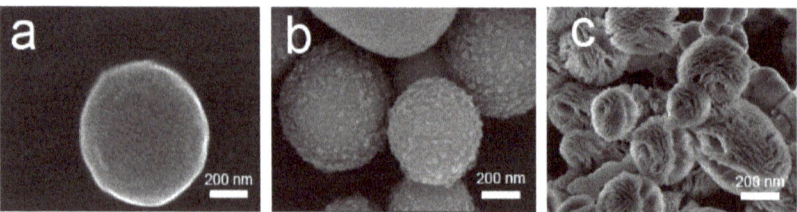

Figure 4. SEM images of CeO_2 samples prepared at 270 °C for different reaction times: (**a**) 6 h; (**b**) 12 h; and (**c**) 24 h.

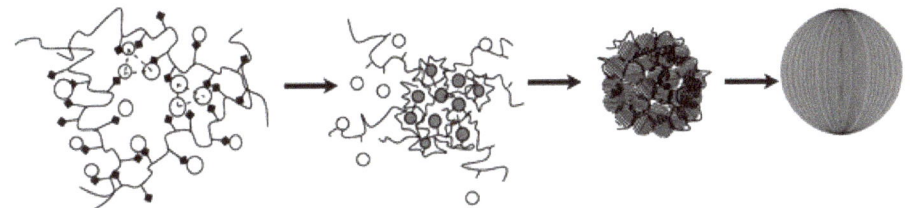

Figure 5. Schematic illustrating the formation of 3D hierarchical structured CeO_2.

3.3. Catalytic Performance of 3D Hierarchical Structured CeO_2 for CO Combustion

Catalytic application is an important direction for CeO_2 researches because the oxygen storage capacity of CeO_2 is associated with its ability to undergo a facile conversion between Ce(III) and Ce(IV). Therefore, the catalytic activity of the as-prepared 3D hierarchical structured CeO_2 was tested by CO oxidation. As shown in Figure 6, the 3D hierarchical structured CeO_2 exhibits better activity toward CO oxidation than commercial CeO_2 (purchased from Beijing Yili Chemical Reagent Co. Ltd., Beijing, China) does. The 50% conversion temperature of the 3D hierarchical structured CeO_2 is about 320 °C, while that of the commercial CeO_2 is higher than 380 °C.

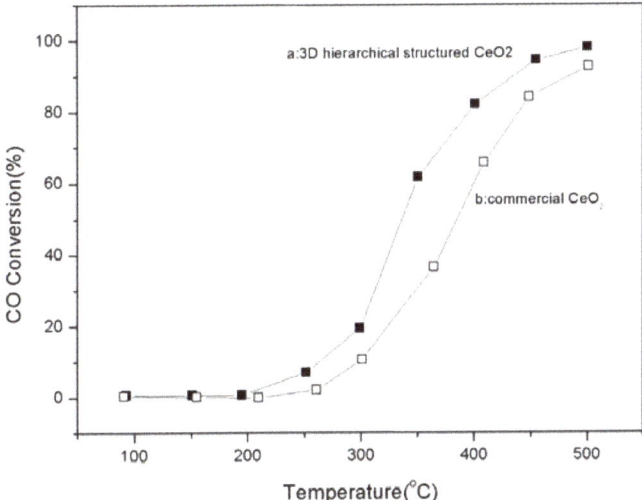

Figure 6. CO conversion rate in the presence of (**a**) as-prepared 3D hierarchical structured CeO_2, and (**b**) commercial CeO_2.

The sample was further characterized by XPS and the Ce 3d electron core level spectra are shown in Figure 7. The four main $3d_{5/2}$ features at 882.7, 885.2, 888.5, and 898.3 eV correspond to V, V', V", and V''' components, respectively. The $3d_{3/2}$ features at 901.3, 903.4, 907.3, and 916.9 eV correspond to U, U', U", and U''' components [15], respectively. The signals V' and U' are characteristic of Ce(III) [16]. According to the ratio of the area for Ce^{3+} peaks to the whole peak area in Ce 3d region, the amount of Ce^{3+} of 3D hierarchical structured CeO_2 is 51.8%. The amount of Ce^{3+} of commercial CeO_2 is 13.2%. The 3D hierarchical structured CeO_2 has a much higher Ce^{3+} concentration, which implies a much higher concentration of oxygen defects compared with commercial CeO_2. A large amount of oxygen defects enhances the conversion between Ce(III) and Ce(IV), thereby supplying more reactive oxygen. Thus, the special structure of 3D hierarchical structured CeO_2 provides more active sites for CO oxidation.

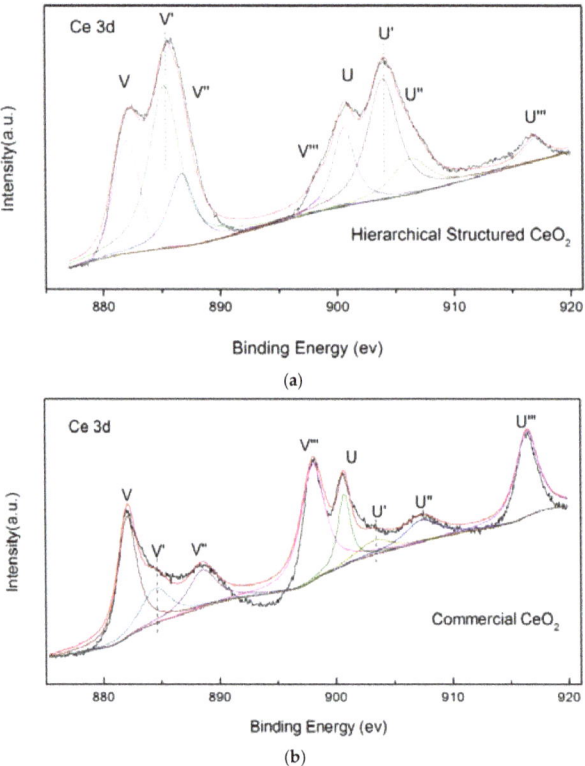

Figure 7. X-ray photoelectron spectra of Ce 3D regions of 3D (**a**) hierarchical structured CeO_2 and (**b**) commercial CeO_2.

4. Conclusions

In summary, a simple and economical hydrothermal route was presented to synthesize 3D hierarchical structured CeO_2 using $CeCl_3 \cdot 7H_2O$ and PVP. The 3D hierarchical structured CeO_2 has a beautiful flower-like structure, which consists of many nanosheets. A two-stage growth process was identified for the formation of 3D hierarchical structured CeO_2, and Ostwald ripening was found to play an important role in the transformation of the nanoparticles into nanosheets. The 3D hierarchical structured CeO_2 exhibited a higher catalytic activity toward CO oxidation compared with commercial CeO_2.

Author Contributions: Conceptualization: G.S. and Q.W.; methodology: G.S.; software: Z.W.; validation: G.S., Q.W.; formal analysis: G.S.; investigation: G.S.; resources: Q.W.; data curation: G.S.; writing—original draft preparation: G.S.; writing—review and editing: M.L.; visualization, M.L.; supervision: Q.W.; project administration: G.S.; funding acquisition: G.S.

Funding: This work was supported financially by the National Natural Science Foundation of China (no. 51402062).

Conflicts of Interest: The authors declare no conflict of interest.

References

1. Cai, L.; Chen, S.H.; Zhao, M.; Gong, M.C.; Shi, Z.H.; Chen, Y.Q. Pd supported three-way catalyst: Preparation of CeO_2-ZrO_2-BaO support and catalytic performance. *Chin. J. Inorg. Chem.* **2009**, *25*, 474–479.
2. Shimazu, M.; Isobe, T.; Ando, S.; Hiwatashi, K.; Ueno, A.; Yamaji, K.; Kishimoto, H.; Yokokawa, H.; Nakajima, A.; Okada, K. Stability of Sc_2O_3 and CeO_2 co-doped ZrO_2 electrolyte during the operation of solid oxide fuel cells. *Solid State Ion.* **2011**, *182*, 120–126. [CrossRef]
3. Lee, D.H.; Cha, K.S.; Lee, Y.S.; Kang, K.S.; Park, C.S.; Kim, Y.H. Effects of CeO_2 additive on redox characteristics of Fe-based mixed oxide mediums for storage and production of hydrogen. *Int. J. Hydrogen Energy* **2009**, *34*, 1417–1422. [CrossRef]
4. Sanghavi, R.; Nandasiri, M.; Kuchibhatla, S.; Jiang, W.L.; Varga, T.; Nachimuthu, P.; Engelhard, M.H.; Shutthanandan, V.; Thevuthasan, S.; Kayani, A.; et al. Thickness dependency of thin-film samaria-doped ceria for oxygen sensing. *IEEE Sens. J.* **2011**, *11*, 217–224. [CrossRef]
5. Lykaki, M.; Pachatouridou, E.; Iliopoulou, E.; Carabineiro, S.A.C.; Konsolakis, M. Impact of the synthesis parameters on the solid state properties and the CO oxidation performance of ceria nanoparticles. *RSC Adv.* **2017**, *7*, 6160–6169. [CrossRef]
6. Lykaki, M.; Pachatouridou, E.; Carabineiro, S.A.C.; Iliopoulou, E.; Andriopoulou, C.; Kallithrakas-Kontos, N.; Boghosian, S.; Konsolakis, M. Ceria nanoparticles shape effects on the structural defects and surface chemistry: Implications in CO oxidation by Cu/CeO_2 catalysts. *Appl. Catal. B Environ.* **2018**, *230*, 18–28. [CrossRef]
7. Meng, F.M.; Lu, F.; Wang, L.N.; Cui, J.B.; Lü, J.G. Novel fabrication and synthetic mechanism of CeO_2 nanorods by a chloride-assisted hydrothermal method. *Sci. Adv. Mater.* **2012**, *4*, 1018–1023. [CrossRef]
8. Deus, R.C.; Cilense, M.; Foschini, C.R.; Ramirez, M.A.; Longo, E.; Simões, A.Z. Influence of mineralizer agents on the growth of crystalline CeO_2 nanospheres by the microwave-hydrothermal method. *J. Alloys Compd.* **2013**, *550*, 245–251. [CrossRef]
9. Zhao, X.B.; You, J.; Lu, X.W.; Chen, Z.G. Hydrothermal synthesis, characterization and property of CeO_2 nanotube. *J. Inorg. Mater.* **2011**, *26*, 159–164. [CrossRef]
10. He, L.A.; Yu, Y.B.; Zhang, C.B.; He, H. Complete catalytic oxidation of o-xylene over CeO_2 nanocubes. *J. Environ. Sci. China* **2011**, *23*, 60–165. [CrossRef]
11. Cao, C.Y.; Cui, Z.M.; Chen, C.Q.; Song, W.G.; Cai, W. Ceria hollow nanospheres produced by a template-free microwave-assisted hydrothermal method for heavy metal ion removal and catalysis. *J. Phys. Chem. C* **2010**, *114*, 9865–9870. [CrossRef]
12. Zhong, L.S.; Hu, J.S.; Cao, A.M.; Liu, Q.; Song, W.G.; Wan, L.J. 3D flowerlike ceria micro/nanocomposite structure and its application for water treatment and CO removal. *Chem. Mater.* **2007**, *19*, 1648–1655. [CrossRef]
13. Li, H.F.; Lu, G.Z.; Dai, Q.G.; Wang, Y.Q.; Guo, Y.; Guo, Y.L. Hierarchical organization and catalytic activity of high-surface-area mesoporous ceria microspheres prepared via hydrothermal routes. *ACS Appl. Mater. Interfaces* **2010**, *2*, 838–846. [CrossRef] [PubMed]
14. Ouyang, X.; Li, W.; Xie, S.; Zhai, T.; Yu, M.; Gan, J.; Lu, X. Hierarchical CeO_2 nanospheres as highly-efficient adsorbents for dye removal. *New J. Chem.* **2013**, *37*, 585–588. [CrossRef]
15. Bêche, E.; Charvin, P.; Perarnau, D.; Abanades, S.; Flamant, G. Ce 3d XPS investigation of cerium oxides and mixed cerium oxide ($Ce_xTi_yO_z$). *Surf. Interface Anal.* **2008**, *40*, 264–267. [CrossRef]
16. Natile, M.M.; Glisenti, A. CoO_x/CeO_2 nanocomposite powders: Synthesis, characterization, and reactivity. *Chem. Mater.* **2005**, *17*, 3403–3414. [CrossRef]

© 2018 by the authors. Licensee MDPI, Basel, Switzerland. This article is an open access article distributed under the terms and conditions of the Creative Commons Attribution (CC BY) license (http://creativecommons.org/licenses/by/4.0/).

Article

Tailoring the Grain Size of Bi-Layer Graphene by Pulsed Laser Deposition

Jin Wang [1,2], Xuemin Wang [2], Jian Yu [2], Tingting Xiao [2], Liping Peng [2], Long Fan [2], Chuanbin Wang [1], Qiang Shen [1,*] and Weidong Wu [2,3,*]

1. State Key Lab of Advanced Technology for Materials Synthesis and Processing, Wuhan University of Technology, Wuhan 430070, China; swustwj@163.com (J.W.); wangcb@whut.edu.cn (C.W.)
2. Science and Technology on Plasma Physics Laboratory, Research Center of Laser Fusion, China Academy of Engineering Physics, Mianyang 621900, China; wangxuemin75@sina.com (X.W.); yujianroy@163.com (J.Y.); tingtingxiao@yeah.net (T.X.); pengliping2005@126.com (L.P.); sfanlong@163.com (L.F.)
3. Collaborative Innovation Center of IFSA (CICIFSA), Shanghai Jiao Tong University, Shanghai 200240, China
* Correspondence: sqqf@whut.edu.cn (Q.S.); wuweidongding@163.com (W.W.)

Received: 11 October 2018; Accepted: 25 October 2018; Published: 1 November 2018

Abstract: Improving the thermoelectric efficiency of a material requires a suitable ratio between electrical and thermal conductivity. Nanostructured graphene provides a possible route to improving thermoelectric efficiency. Bi-layer graphene was successfully prepared using pulsed laser deposition in this study. The size of graphene grains was controlled by adjusting the number of pulses. Raman spectra indicated that the graphene was bi-layer. Scanning electron microscopy (SEM) images clearly show that graphene changes from nanostructured to continuous films when more pulses are used during fabrication. Those results indicate that the size of the grains can be controlled between 39 and 182 nm. A detailed analysis of X-ray photoelectron spectra reveals that the sp^2 hybrid state is the main chemical state in carbon. The mobility is significantly affected by the grain size in graphene, and there exists a relatively stable region between 500 and 800 pulses. The observed phenomena originate from competition between decreasing resistance and increasing carrier concentration. These studies should be valuable for regulating grains sizes for thermoelectric applications of graphene.

Keywords: graphene; PLD; mobility

1. Introduction

With increasingly serious environmental pollution and an energy crisis, it is very important to reduce environmental pollution and convert waste heat into electrical energy. For this reason, it is necessary to find efficient thermoelectric conversion materials. Excellent thermoelectric efficiency requires high electrical conductivity and low thermal conductivity. Nanostructured materials [1] limit the mean free path of electrons while restricting heat conduction. This shows that the electrical properties of nanomaterials are related to their special structures [2–4]. Nanostructured graphene has special electrical transport properties and is expected to have high thermoelectric efficiency [5–8]. Previous studies show that nanostructured graphene can provide significantly reduced thermal conductivity with little effect on electrical conductivity [9]. Thus, nanostructured graphene with controllable grain size can greatly improve the thermoelectric efficiency. Currently, the mainstream method for preparing graphene is chemical vapor deposition (CVD) [10–13]. Most researchers focus on the properties of single grain graphene, but the influence of crystal grain size on electrical conductivity of graphene is still unclear at the macroscopic scale [14–17]. The primary reasons for those observed phenomena originate from the fact that it is difficult to use CVD methods to adjust the size of graphene nanograins. Therefore, the preparation of graphene with controllable grain size is the key to expanding the applications of graphene, especially thermoelectric applications [18,19]. Because pulsed laser deposition (PLD) can be used to controllably generate highly energetic carbon species [18–22], it has natural advantages in controlling graphene crystal grains. This method is suitable for adjusting the size of graphene grains. Early experiments examined the effects of laser energy, substrate temperature, ablation time, and cooling rate [23–27]. However, research on the control of graphene crystal grains by PLD is still deficient.

Bi-layer graphene was prepared using PLD in this study. The effect of pulse numbers on the size of graphene grains was studied. In this case, the growth process of bi-layer graphene grains could be sufficiently controlled.

2. Experimental

Graphene grains were deposited on single crystal Cu (111) substrates by PLD. An excimer KrF laser was used for ablation. The specific experimental parameters are listed in Table 1. The number of pulses was set to 300, 500, 700, 800, and 900, and the corresponding samples are labeled in Table 2. Raman spectra from the graphene samples were gathered using a 514-nm laser in backscattering geometry at room temperature (Invia, Renishaw, London, UK). A field emission scanning electron microscopy (FE-SEM) (Quanta 250, FEI, Hillsboro, OR, USA) operated at 15 kV was used to examine the surface morphology of graphene and view the graphene grains. The working distance was 5 mm. The grain size distribution and average grain size of graphene were calculated using Nano Measurer software (Nano Measurer v1.2.5). We measured graphene grains with different sizes and in different regions in the SEM images in order to ensure accurate results. X-ray photoelectron spectra (XPS) spectra were gathered with an ESCALAB 250Xi XPS while the samples were excited with Al Kα radiation. The electrical properties of graphene were determined from Hall measurements.

Table 1. Experimental fabrication parameters.

Experiment Conditions	Experimental Parameters
Background vacuum	2.0×10^{-6} Pa
Working vacuum	4.5×10^{-5} Pa
Target	highly oriented pyrolytic graphite (HOPG) (purity > 99.99%)
Substrate	single crystal Cu (111)
Laser pulse frequency	1 Hz
Energy density	4 J/cm^2
Distance between the target and the substrate	10 cm
Annealing condition	1000 °C

Table 2. Sample numbers and their corresponding pulse numbers.

Samples	1$^{\#}$	2$^{\#}$	3$^{\#}$	4$^{\#}$	5$^{\#}$
Number of pulses	300	500	700	800	900

3. Results and Discussion

Figure 1 shows the Raman spectra from the graphene films deposited on the single crystal Cu (111) surface. Four peaks are present in the Raman spectra. The strong D peak at 1350 cm^{-1} was induced by disorder in the atomic arrangement, the edge effect of graphene, or ripples and charge puddles. This means a significant number of defects appeared in the graphene thin films. The G peak at approximately 1580 cm^{-1} originates from highly oriented graphite induced by the doubly degenerate zone center E$_{2g}$ mode. The 2D peak at approximately 2700 cm^{-1} originates from the double resonance Raman excitation of two-photon near two mutually nonequivalent K points at the center of the first Brillouin zone. The intensities of I_G, I_{2D}, and their ratios are useful indicators of the quality and number of layers in the graphene samples. The specific peak information, I_D/I_G, and I_{2D}/I_G ratios are shown in Table 3. As the number of pulses increases, the peak intensity of the D, G, and 2D peaks in the Raman spectra constantly increased. The ratio of I_{2D}/I_G fell in between 0.79 and 0.94, implying the graphene layers have a bi-layer structure [28–31]. This means that the number of graphene layers remains constant as the number of pulses increases. One possible reason is that single crystal Cu (111) may play a role in limiting or preventing precipitation altogether at 1000 °C [32]. Another peak in Figure 1 at approximately 2960 cm^{-1} (called D + D') is a dual-phonon process peak originating from one intravalley and one intervalley phonon scattering [33]. This peak is closely related to the defect state. The peak intensity increased significantly as the defect density increased. Interestingly, the D + D' peak is only observed in graphene prepared by PLD, and the peak has not been observed in graphene films prepared with other methods.

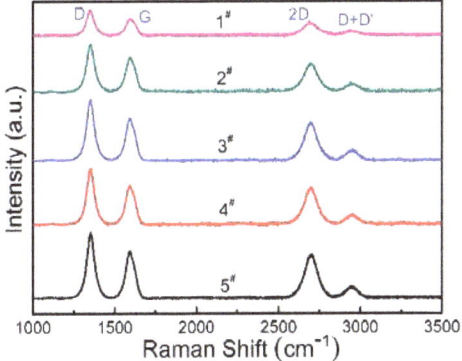

Figure 1. Raman spectra of graphene from samples of 1#–5#.

Table 3. Raman intensity for I_D, I_G, I_{2D}, and the ratio of I_D/I_G and I_{2D}/I_G from Figure 1.

Samples	D-Band Position	D-Band Intensity	G-Band Position	G-Band Intensity	2D-Band Position	2D-Band Intensity	I_D/I_G	I_{2D}/I_G
1#	1349 cm^{-1}	3240	1598 cm^{-1}	2103	2688 cm^{-1}	1728	1.54	0.82
2#	1354 cm^{-1}	5922	1595 cm^{-1}	4333	2690 cm^{-1}	3426	1.37	0.79
3#	1350 cm^{-1}	7572	1592 cm^{-1}	5262	2693 cm^{-1}	4683	1.44	0.89
4#	1354 cm^{-1}	6913	1595 cm^{-1}	4834	2693 cm^{-1}	4528	1.43	0.94
5#	1350 cm^{-1}	8143	1595 cm^{-1}	5849	2704 cm^{-1}	5359	1.39	0.92

The morphology of graphene is clearly shown in the SEM image in Figure 2. Figure 2a–e show SEM images of graphene from samples of 1#–5#, respectively. Sample 1# contains small and discontinuous graphene grains. The corresponding grain size distribution is shown in Figure 3a, where the average graphene grain size is 39 nm. The formation of graphene nanocrystals is caused by multiple nucleation sites on the surface of the Cu (111) substrate at a small number of pulses. The step on the Cu (111) surface results from high temperature. As shown in sample 2#, it was found that graphene nanograins are connected to each other to form graphene grains with an average size of 66 nm when the number of pulses increased to 500. Although the small grains are connected to each other to form larger graphene grains, it can be seen from the figure that there are still many discontinuities. Sample 3# was grown using 700 pulses, resulting in large graphene grains with an average size of approximately 182 nm. In sample 4#, the number of pulses increased to 800. It can be seen that the graphene grains are almost all connected together to form a continuous film in Figure 3e, but there are still discontinuities shown in the white dotted ellipse. In sample 5#, full coverage is achieved, indicating that a completely continuous graphene film was formed. The size of bi-layer graphene grains was controlled by adjusting the number of pulses. Nanostructured graphene prepared by PLD growth gives hope that one would have a much better control of the thermal properties of supported bi-layer graphene since the grain size has an effect on $K(T)$ of graphene. Recently, studies [34–36] show that acoustic flexure (ZA) modes are the dominant heat transport in graphene based on the dependence of $K(T) \sim T^{1.4}$ or $\sim T^{1.5}$. That means K can be adjusted within a certain range by controlling the graphene structure.

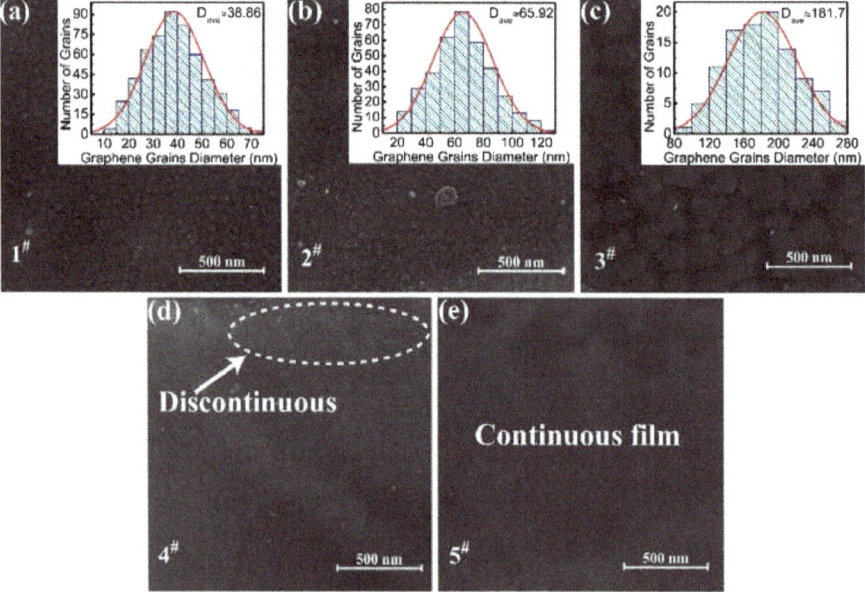

Figure 2. Scanning electron microscopy (SEM) image of (**a**) sample 1#, (**b**) sample 2#, (**c**) sample 3#, (**d**) sample 4#, and (**e**) sample 5#. The inset in (**a**–**c**) show the corresponding grain size distribution. The white dotted ellipse in (**d**) shows the discontinuous part.

XPS measurements [37] can provide direct evidence of the chemical states in graphene. Figure 3 shows the XPS spectra from graphene grown using different numbers of pulses. Figure 3a shows XPS spectra from each sample, which indicate the existence of C, O, and Cu. The main features correspond to C 1s, O 1s, and Cu 2p3. The major species remaining were C=C (284.7 eV). The C 1s spectra from all samples are shown in Figure 3b–f, respectively. Peak A at 284.7 eV (C 1s) is attributed to sp^2 carbon bonds, which agrees with the component of graphene [38]. It is well known that graphene formation occurs due to the surface graphitization of carbon films. Peak B at 285.50.1 eV corresponds to sp^3 carbon atoms. Peak C exhibits much smaller intensity at about 286.3 eV and is attributed to some C–O contamination at the surface of the films due to exposure to air [39]. The XPS results show that the growth kinetic energy provided by the PLD system cannot induce a complete transformation of all sp^3 bonds into sp^2 bonds in graphene. These results indicate the presence of growth defects during graphene preparation using PLD.

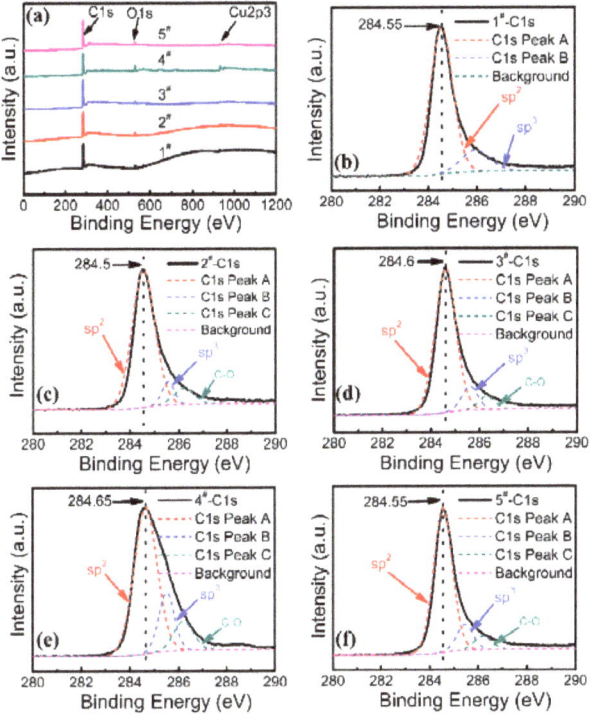

Figure 3. (a) X-ray photoelectron spectra (XPS) spectra from graphene grown using different number of pulses. (b–f) C 1s peaks in graphene from samples $1^{\#}$–$5^{\#}$, respectively.

Figure 4a shows the room-temperature mobility of graphene with different numbers of pulses measured by the Hall effect. This clearly implies that the measured graphene mobility is very low in the experiment. The low mobility of graphene occurs due to grain boundaries and defects. Meanwhile, the mobility of graphene is basically stable, especially when grown using between 500 and 800 pulses. The mobility increases as the number of pulses used during growth decreases. The mobility of graphene is determined using the formula $\mu = \sigma/ne$, where σ is the electrical conductivity [40]. $\sigma = 1/\rho$, where ρ is the resistivity. Therefore, the graphene mobility formula can be simplified as $\mu = 1/R_\Omega \cdot n \cdot e$, where R_Ω is the sheet resistance and n is the carrier concentration. The relationship between R_Ω and n and the number of pulses used during growth was studied in order to better understand its effect on mobility. Figure 4b shows that the carrier concentration n is ~10^{13}, which is one to two orders of magnitude higher than graphene with good mobility [41]. The measured resistance of the graphene samples is several kΩ. The two factors result in low mobility. With the increasing number of pulses, it can be seen from the previous Raman spectra results that the defect peak D gradually increased, indicating that the defect state density in graphene increased. As the density of the defect state increased, n also increased. However, R_Ω decreases with increasing grain size. The mobility μ becomes relatively stable when the competition between R_Ω and n is balanced. This is the reason why the mobility of graphene is basically stable when grown using 500 to 800 pulses.

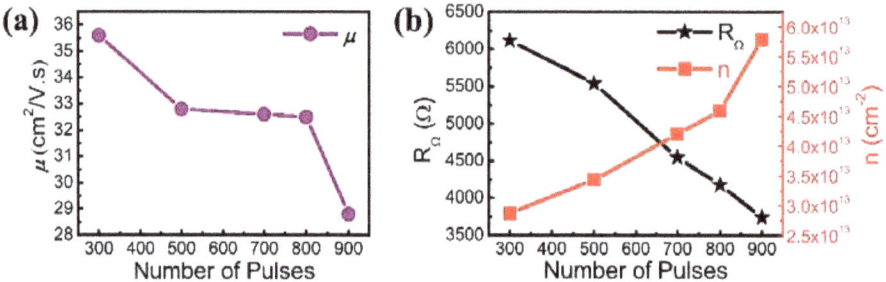

Figure 4. (a) Room-temperature mobility of graphene with different number of pulses; (b) n and R_Ω of graphene with the samples of 1#, 2#, 3#, 4#, and 5#, respectively.

4. Conclusions

In conclusion, we prepared bi-layer graphene from a solid carbon source using PLD. The grain size of graphene can be controlled between 39 and 182 nm by varying the number of pulses from 300 to 900. Regarding the chemical structure, sp^3 bonds exist in graphene, which lead to many defects during graphene growth. Electronic mobility can be affected by grain size and becomes relatively stable between 500 and 800 pulses. These results may occur due to competition between resistance and carrier concentration. These findings can be used to tune the grain size of graphene, and the results are beneficial for thermoelectric applications.

Author Contributions: Data curation, Formal analysis, Writing-original draft, J.W.; Methodology, X.W.; Formal analysis, J.Y.; Data curation, T.X.; Investigation, L.P.; Data Curation, L.F.; Investigation, C.W.; Funding Acquisition, Supervision, Q.S. and W.W.

Funding: This work was financially supported by the National Natural Science Foundation of China (No. 51521001 and 51872217), and the "111" project (No. B13035).

Conflicts of Interest: The authors declare no conflict of interest.

References

1. Zhao, L.D.; Lo, S.H.; Zhang, Y.; Sun, H.; Tan, G.; Uher, C.; Wolverton, C.; Dravid, V.P.; Kanatzidis, M.G. Ultralow thermal conductivity and high thermoelectric figure of merit in SnSe crystals. *Nature* **2014**, *508*, 373–377. [CrossRef] [PubMed]
2. Cavallaro, G.; Chiappisi, L.; Pasbakhsh, P.; Gradzielski, M.; Lazzara, G. A structural comparison of halloysite nanotubes of different origin by Small-Angle Neutron Scattering (SANS) and Electric Birefringence. *Appl. Clay Sci.* **2018**, *160*, 71–80. [CrossRef]
3. Cavallaro, G.; Grillo, I.; Gradzielski, M.; Lazzara, G. Structure of hybrid materials based on halloysite nanotubes filled with anionic surfactants. *J. Phys. Chem. C* **2016**, *120*, 13492–13502. [CrossRef]
4. Lazzara, G.; Cavallaro, G.; Panchal, A.; Fakhrullin, R.; Stavitskaya, A.; Vinokurov, V.; Lvov, Y. An assembly of organic-inorganic composites using halloysite clay nanotubes. *Curr. Opin. Colloid Interface Sci.* **2018**, *35*, 42–50. [CrossRef]
5. Dragoman, D.; Dragoman, M. Giant thermoelectric effect in graphene. *Appl. Phys. Lett.* **2007**, *91*, 203116. [CrossRef]
6. Xu, X.; Gabor, N.M.; Alden, J.S.; van der Zande, A.M.; McEuen, P.L. Photo-thermoelectric effect at a graphene interface junction. *Nano Lett.* **2009**, *10*, 562–566. [CrossRef] [PubMed]
7. Kim, G.H.; Hwang, D.H.; Woo, S.I. Thermoelectric properties of nanocomposite thin films prepared with poly (3, 4-ethylenedioxythiophene) poly (styrenesulfonate) and graphene. *Phys. Chem. Chem. Phys.* **2012**, *14*, 3530–3536. [CrossRef] [PubMed]
8. Shiau, L.L.; Wang, X.; Goh, S.C.K.; Chuan, K.; Ernst, H.; Tay, B.K. First demonstration of gate voltage-less chemical vapour deposition graphene for non-vacuum thermoelectric study. In Image Sensing Technologies: Materials, Devices, Systems, and Applications V. *Int. Soc. Opt. Photonics* **2018**, *10656*, 106561V.

9. Ma, T.; Liu, Z.; Wen, J.X.; Gao, Y.; Ren, X.B.; Chen, H.J.; Jin, C.H.; Ma, X.L.; Xu, N.S.; Cheng, H.M.; et al. Tailoring the thermal and electrical transport properties of graphene films by grain size engineering. *Nat. Commun.* **2017**, *8*, 14486. [CrossRef] [PubMed]
10. Yu, Q.; Lian, J.; Siriponglert, S.; Li, H.; Chen, Y.P.; Pei, S.S. Graphene segregated on Ni surfaces and transferred to insulators. *Appl. Phys. Lett.* **2008**, *93*, 113103. [CrossRef]
11. Liu, W.; Li, H.; Xu, C.; Khatami, Y.; Banerjee, K. Synthesis of high-quality monolayer and bilayer graphene on copper using chemical vapor deposition. *Carbon* **2011**, *49*, 4122–4130. [CrossRef]
12. Kim, K.S.; Zhao, Y.; Jang, H.; Lee, S.Y.; Kim, J.M.; Kim, K.S.; Ahn, J.H.; Kim, P.; Choi, J.Y.; Hong, B.H. Large-scale pattern growth of graphene films for stretchable transparent electrodes. *Nature* **2009**, *457*, 706–710. [CrossRef] [PubMed]
13. Ago, H.; Ito, Y.; Mizuta, N.; Yoshida, K.; Hu, B.; Orofeo, C.M.; Tsuji, M.; Ikeda, K.; Mizuno, S. Epitaxial chemical vapor deposition growth of single-layer graphene over cobalt film crystallized on sapphire. *ACS Nano* **2010**, *4*, 7407–7414. [CrossRef] [PubMed]
14. Perez, S.B.; Balbuena, P.B. Formation of Multilayer Graphene Domains with Strong Sulfur-Carbon Interaction and Enhanced Sulfur Reduction Zones for Lithium-Sulfur Battery Cathodes. *ChemSusChem* **2018**, *11*, 1970–1980. [CrossRef] [PubMed]
15. Yazyev, O.V.; Chen, Y.P. Polycrystalline graphene and other two-dimensional materials. *Nat. Nanotechnol.* **2014**, *9*, 755–767. [CrossRef] [PubMed]
16. Fei, Z.; Rodin, A.S.; Gannett, W.; Dai, S.; Regan, W.; Wagner, M.; Liu, M.K.; McLeod, A.S.; Dominguez, G.; Thiemens, M.; et al. Electronic and plasmonic phenomena at graphene grain boundaries. *Nat. Nanotechnol.* **2013**, *8*, 821–825. [CrossRef] [PubMed]
17. Cummings, A.W.; Duong, D.L.; Nguyen, V.L.; Van Tuan, D.; Kotakoski, J.; Barrios Vargas, J.E.; Lee, Y.H.; Roche, S. Charge transport in polycrystalline graphene: challenges and opportunities. *Adv. Mater.* **2014**, *26*, 5079–5094. [CrossRef] [PubMed]
18. Cappelli, E.; Iacobucci, S.; Scilletta, C.; Flammini, R.; Orlando, S.; Mattei, G.; Ascarelli, P.; Borgatti, F.; Giglia, A.; Mahne, N.; Nannarone, S. Orientation tendency of PLD carbon films as a function of substrate temperature: A NEXAFS study. *Diam. Relat. Mater.* **2005**, *14*, 959–964. [CrossRef]
19. Scilletta, C.; Servidori, M.; Orlando, S.; Cappelli, E.; Barba, L.; Ascarelli, P. Influence of substrate temperature and atmosphere on nano-graphene formation and texturing of pulsed Nd: YAG laser-deposited carbon films. *Appl. Surf. Sci.* **2006**, *252*, 4877–4881. [CrossRef]
20. Cappelli, E.; Orlando, S.; Servidori, M.; Scilletta, C. Nano-graphene structures deposited by N-IR pulsed laser ablation of graphite on Si. *Appl. Surf. Sci.* **2007**, *254*, 1273–1278. [CrossRef]
21. Xiong, Z.W.; Cao, L.H. Interparticle spacing dependence of magnetic anisotropy and dipolar interaction of Ni nanocrystals embedded in epitaxial $BaTiO_3$ matrix. *Ceram. Int.* **2018**, *44*, 8155–8160. [CrossRef]
22. Xiong, Z.W.; Cao, L.H. Red-ultraviolet photoluminescence tuning by Ni nanocrystals in epitaxial $SrTiO_3$ matrix. *Appl. Surf. Sci.* **2018**, *445*, 65–70. [CrossRef]
23. Kumar, I.; Khare, A. Multi-and few-layer graphene on insulating substrate via pulsed laser deposition technique. *Appl. Surf. Sci.* **2014**, *317*, 1004–1009. [CrossRef]
24. Xu, S.C.; Man, B.Y.; Jiang, S.Z.; Liu, A.H.; Hu, G.D.; Chen, C.S.; Liu, M.; Yang, C.; Feng, D.J.; Zhang, C. Direct synthesis of graphene on any nonmetallic substrate based on KrF laser ablation of ordered pyrolytic graphite. *Laser Phys. Lett.* **2014**, *11*, 096001. [CrossRef]
25. Dong, X.M.; Liu, S.B.; Song, H.Y.; Gu, P.; Li, X.L. Few-layer graphene film fabricated by femtosecond pulse laser deposition without catalytic layers. *Chin. Opt. Lett.* **2015**, *13*, 021601. [CrossRef]
26. Na, B.J.; Kim, T.H.; Lee, C.; Lee, S.H. Study on Graphene Thin Films Grown on Single Crystal Sapphire Substrates Without a Catalytic Metal Using Pulsed Laser Deposition. *Trans. Electr. Electron. Mater.* **2015**, *16*, 70–73. [CrossRef]
27. Koh, A.T.; Foong, Y.M.; Chua, D.H. Comparison of the mechanism of low defect few-layer graphene fabricated on different metals by pulsed laser deposition. *Diam. Relat. Mater.* **2012**, *25*, 98–102. [CrossRef]
28. Li, X.S.; Cai, W.W.; An, J.; Kim, S.; Nah, J.; Yang, D.X.; Piner, R.; Velamakanni, A.; Jung, I.; Tutuc, E.; et al. Large-area synthesis of high-quality and uniform graphene films on copper foils. *Science* **2009**, *324*, 1312–1314. [CrossRef] [PubMed]

29. Reina, A.; Jia, X.T.; Ho, J.; Nezich, D.; Son, H.; Bulovic, V.; Dresselhaus, M.S.; Kong, J. Large area, few-layer graphene films on arbitrary substrates by chemical vapor deposition. *Nano Lett.* **2008**, *9*, 30–35. [CrossRef] [PubMed]
30. Cao, H.L.; Yu, Q.K.; Colby, R.; Pandey, D.; Park, C.S.; Lian, J.; Zemlyanov, D.; Childres, I.; Drachev, V.; Stach, E.A.; et al. Large-scale graphitic thin films synthesized on Ni and transferred to insulators: Structural and electronic properties. *J. Appl. Phys.* **2010**, *107*, 044310. [CrossRef]
31. Abd Elhamid, A.E.M.; Hafez, M.A.; Aboulfotouh, A.M.; Azzouz, I.M. Study of graphene growth on copper foil by pulsed laser deposition at reduced temperature. *J. Appl. Phys.* **2017**, *121*, 025303. [CrossRef]
32. Yazyev, O.V.; Pasquarello, A. Effect of metal elements in catalytic growth of carbon nanotubes. *Phys. Rev. Lett.* **2008**, *100*, 156102. [CrossRef] [PubMed]
33. Ferrari, A.C.; Basko, D.M. Raman spectroscopy as a versatile tool for studying the properties of graphene. *Nat. Nanotechnol.* **2013**, *8*, 235–246. [CrossRef] [PubMed]
34. Xu, X.; Wang, Y.; Zhang, K.; Zhao, X.; Bae, S.; Heinrich, M.; Bui, C.T.; Xie, R.; Thong, J.T.L.; Hong, B.H.; et al. Phonon transport in suspended single layer graphene. *arXiv* **2010**.
35. Wang, Z.Q.; Xie, R.G.; Bui, C.T.; Liu, D.; Ni, X.X.; Li, B.W.; Thong, J.T. Thermal transport in suspended and supported few-layer graphene. *Nano Lett.* **2011**, *11*, 113–118. [CrossRef] [PubMed]
36. Pettes, M.T.; Jo, I.; Yao, Z.; Shi, L. Influence of polymeric residue on the thermal conductivity of suspended bilayer graphene. *Nano Lett.* **2011**, *11*, 1195–1200. [CrossRef] [PubMed]
37. Yu, J.; Xiao, T.T.; Wang, X.M.; Zhao, Y.; Li, X.J.; Xu, X.B.; Xiong, Z.W.; Wang, X.M.; Peng, L.P.; Wang, J.; et al. Splitting of the ultraviolet plasmon resonance from controlling FePt nanoparticles morphology. *Appl. Surf. Sci.* **2018**, *435*, 1–6. [CrossRef]
38. Pirkle, A.; Chan, J.; Venugopal, A.; Hinojos, D.; Magnuson, C.W.; McDonnell, S.; Colombo, L.; Vogel, E.M.; Ruoff, R.S.; Wallace, R.M. The effect of chemical residues on the physical and electrical properties of chemical vapor deposited graphene transferred to SiO$_2$. *Appl. Phys. Lett.* **2011**, *99*, 122108. [CrossRef]
39. Siokou, A.; Ravani, F.; Karakalos, S.; Frank, O.; Kalbac, M.; Galiotis, C. Surface refinement and electronic properties of graphene layers grown on copper substrate: an XPS, UPS and EELS study. *Appl. Surf. Sci.* **2011**, *257*, 9785–9790. [CrossRef]
40. Liu, L.; Chen, J.J.; Zhou, Z.G.; Yi, Z.; Ye, X. Tunable absorption enhancement in electric split-ring resonators-shaped graphene arrays. *Mater. Res. Express* **2018**, *5*, 045802. [CrossRef]
41. Morozov, S.V.; Novoselov, K.S.; Katsnelson, M.I.; Schedin, F.; Elias, D.C.; Jaszczak, J.A.; Geim, A.K. Giant intrinsic carrier mobilities in graphene and its bilayer. *Phys. Rev. Lett.* **2008**, *100*, 016602. [CrossRef] [PubMed]

© 2018 by the authors. Licensee MDPI, Basel, Switzerland. This article is an open access article distributed under the terms and conditions of the Creative Commons Attribution (CC BY) license (http://creativecommons.org/licenses/by/4.0/).

Article

Application of Glycation in Regulating the Heat-Induced Nanoparticles of Egg White Protein

Chenying Wang [1], Xidong Ren [2,3], Yujie Su [1,*] and Yanjun Yang [1,*]

1. State Key Laboratory of Food Science and Technology and School of Food Science and Technology, Jiangnan University, Wuxi 214122, China; chenying071776@163.com
2. State Key Laboratory of Biobased Material and Green Papermaking, Qilu University of Technology, Shandong Academy of Sciences, Jinan 250353, China; renxidong1986@126.com
3. Shandong Provincial Key Laboratory of Microbial Engineering, Department of Bioengineering, Qilu University of Technology, Shandong Academy of Sciences, Jinan 250353, China
* Correspondence: suyujie@jiangnan.edu.cn (Y.S.); yangyj@jiangnan.edu.cn (Y.Y.); Tel.: +86-510-85329080

Received: 24 October 2018; Accepted: 14 November 2018; Published: 15 November 2018

Abstract: Due to the poor thermal stability of egg white protein (EWP), important challenges remain regarding preparation of nanoparticles for EWP above the denaturation temperature at neutral conditions. In this study, nanoparticles were fabricated from conjugates of EWP and isomalto-oligosaccharide (IMO) after heating at 90 °C for 30 min. Meanwhile, the effects of protein concentration, temperature, pH, ionic strength and degree of glycation (DG) on the formation of nanoparticles from IMO-EWP were investigated. To further reveal the formation mechanism of the nanoparticles, structures, thermal denaturation properties and surface properties were compared between EWP and IMO-EWP conjugates. Furthermore, the emulsifying activity index (EAI) and the emulsifying stability index (ESI) of nanoparticles were determined. The results indicated that glycation enhanced thermal stability and net surface charge of EWP due to changes in the EWP structure. The thermal aggregation of EWP was inhibited significantly by glycation, and enhanced with a higher degree of glycation. Meanwhile, the nanoparticles (<200 nm in size) were obtained at pH 3.0, 7.0 and 9.0 in the presence of NaCl. The increased thermal stability and surface net negative charge after glycation contributed to the inhibition. The EAI and ESI of nanoparticles were increased nearly 3-fold and 2-fold respectively, as compared to unheated EWP.

Keywords: egg white protein; isomalto-oligosaccharide; glycation; thermal aggregation; nanoparticle; emulsifying property

1. Introduction

Egg white protein (EWP) is an important ingredient in food processing, because of its abundant nutritive value and various functional properties. The main components of EWP are ovalbumin (OVA, 54%), ovotransferrin (OT, 12%), ovomucoid (OM, 11%) and lysozyme (LY, 3.4%) [1]. These proteins mainly show a globular structure. However, its industrial application is limited by its poor thermal stability and emulsifying properties. It was reported that heat-induced nanoparticles formed from globular proteins could increase their emulsifying capacity and binding ability to hydrophobic bioactive compounds [2–4]. Although half of the amino acid residues in OVA are hydrophobic, EWP presents good water-solubility, since most hydrophobic amino acid residues are embedded into protein molecules under natural conditions. Heating promotes EWP unfolding, in which hydrophobic amino acids are exposed, conferring an increase in protein surface hydrophobicity [5]. However, considering the thermolability of EWP, and in particular the OT (denaturation temperature TD = 62 °C, pI = 6.1) [6], the aggregation of which occurs easily under neutral conditions, the application of thermal modification technology in improving the functional properties of EWP is limited. Attempts in fabrication of EWP nanoparticles by heat treatment under extremely acidic [7] or alkaline conditions [8] have been successful. This proves that the heat-induced EWP nanoparticles exhibit excellent potential to be a type of delivery system for hydrophobic compounds. However, the application systems for extremely acidic or alkaline conditions are limited. Therefore, the preparation of nanoparticles suitable for neutral systems is very important. However, the stable heat-induced nanoparticle prepared by commercial EWP at neutral condition was not reported.

Glycation by Maillard reaction (MR) between reducing sugar and free amino group in proteins to form cross-links are one of the few chemical modification methods applicable to food production. Many studies have succeeded in suppressing thermal aggregation of proteins using glycation, such as whey protein [9], soy protein [10], peanut protein [11] and superoxide dismutase [12]. In addition, the MR would partially unfold the EWP molecules and expose the hydrophobic group to the surface, further improving the oil-in-water emulsifying ability [13]. Therefore, preparation of heat-induced nanoparticles from glycated EWP may be more beneficial for the emulsion system.

In this study, the isomalto-oligosaccharide (IMO) and EWP conjugates (IMO-EWP) were prepared by dry-heating MR. Sodium dodecyl sulfate-polyacrylamide gel electrophoresis (SDS-PAGE) and Fourier transform infrared spectroscopy (FTIR) were first used to characterize the structural changes of EWP after glycation. Then, a comparative study of thermal denaturation properties, surface hydrophobicity and ζ-potential were conducted at neutral conditions. The effects of temperature, pH and ionic strength on the thermal aggregation of IMO-EWP dispersions were further investigated. Finally, emulsifying properties were examined to verify the functional properties of fabricated nanoparticles. This study is expected to provide useful information for the preparation of EWP nanoparticles and expand the application of EWP in emulsion systems.

2. Materials and Methods

2.1. Materials

Hen EWP powder was provided by Rongda Co., Ltd. (Xuancheng, China). The EWP powder was manufactured from fresh egg white after removing glucose, followed by spray-drying Isomalto-oligosaccharide-900 (IMO) with an average molecular weight of 564 Da and reducing sugar content of 21.52% was purchased from Baiyou Bio-Technology Co., Ltd. (Langfang, China). The composition of IMO (Figure S1 and Table S1) is shown in Supplementary Materials. Other reagents were purchased from Sinopharm Chemical Reagent Co., Ltd. (Shanghai, China).

2.2. Preparation of IMO-EWP Conjugates

IMO and EWP were separately mixed together at weight ratios of 1:2, 1:10 and 1:40 (the IMO ratios were also indicated as the weight ratio of IMO to protein of 2.5%, 10% and 50%) and dissolved in distilled water, then adjusted to pH 7.0 with 0.1 M NaOH and 0.1 M HCl. After centrifuging (6000× g, 30 min) and filtering, the supernatant was collected and freeze-dried. The freeze-dried EWP without IMO addition was prepared the same as the mixtures. The resulting freeze-dried EWP and mixtures were then placed at 60 °C for 3 days at a relative humidity of 79%. As a control, EWP and IMO were individually maintained for 3 days under the same conditions. The samples were dissolved and dialyzed (molecular mass cut off 3.5 kDa) against distilled water for 3 days at 4 °C to remove the unreacted IMOs. After freeze-drying, the samples were stored at 4°C before further experimentation. The heated conjugates were designated as 2.5% IMO-EWP, 10% IMO-EWP and 50% IMO-EWP, respectively. The protein content of freeze-dried EWP and conjugates were measured via Kjeldahl method.

2.3. Sodium Dodecyl Sulfate-Polyacrylamide Gel Electrophoresis (SDS-PAGE)

Reducing SDS-PAGE was performed on slab gels (12% separating gel and 5% stacking gel) [13]. Samples were dissolved or diluted to a protein concentration of 2.5 mg/mL. The sample supernatants after centrifugation were added into each lane after mixing with sample loading buffer. 0.2% Coomassie Brilliant Blue G-250 and 0.5% periodic acid fuchsin (PAS) were used to stain the protein and carbohydrates in the gel, respectively. Bovine albumin was used as negative control for PSA glycoprotein staining test.

2.4. Grafting Degree (DG) of the IMO-EWP Conjugates

O-phthalaldehyde (OPA) method [14] was employed to measure the free amino groups of proteins. Two hundred microlitres of sample were mixed with 4 mL of OPA solution. Distilled water was used as blank. The absorbances were measured at 340 nm.

The DG of native EWP and IMO-EWP conjugates were calculated as follows [13]:

$$DG\% = (A_0 - A_c)/A_0 \tag{1}$$

where A_0 and A_c are free amino groups content of native EWP and conjugates, respectively.

2.5. Fourier Transform Infrared Spectroscopy (FTIR)

The secondary structural changes of conjugates were analyzed by FTIR. An FTIR IS10 spectrometer (Nicolet Co., Madison, WI, USA) was used to determine the FTIR spectra of native and glycated EWPs that were previously deposited on infrared-transparent sodium bromide (KBr) windows. The FTIR spectra were measured from 4000 to 400 cm^{-1} for 16 scans. After being baseline corrected, and the area was normalized between 1600 and 1710 cm^{-1} using the PeakFit v4.12 (SeaSolve, Framingham, MA, USA). Quantitative estimation of secondary structure components was performed using Gaussian peaks and curve-fitting models [15].

2.6. Differential Scanning Calorimetry (DSC)

The thermal property was measured on a TA Q2000-DSC thermal analyzer (TA Instruments, New Castle, PA, USA) according to the method previously described by Liu [11] with some modification. Samples (protein concentration of 35%, w/v; pH 7.0) were heated from 30–100 °C at a linear rate of 5 °C/min. TA Universal Analysis 2000 was used to calculate enthalpy changes of denaturation (ΔH) and the denaturation temperature (T_d).

2.7. Measurement of Surface Hydrophobicity (H_o)

Samples were dissolved in distilled water to a soluble protein concentration of 10 mg/mL. Then the filtered protein solutions were diluted with 10 mM phosphate buffer (pH 7.0) to a series of five protein concentrations ranging from 0.005–0.02% w/w. The surface hydrophobicity of treated samples was determined using the fluorescence probe of 1-anilino-8-naphthalenesulfonate (ANS) [16]. The fluorescence intensity (FI) was measured with an F-7000 spectrofluorometer (Hitachi, Tokyo, Japan). The initial slope of the FI-protein concentration (mg/mL) plots was used as an index of H_o.

2.8. Preparation of Heat-Induced Aggregate Particles

EWP and IMO-EWP conjugates were separately dissolved in distilled water to corresponding soluble protein concentrations (1–5%, w/v). Then the solutions were adjusted to pH 3.0–9.0 with 0.5 M NaOH and HCl, and 0–150 mM NaCl was added. All protein solutions were filtered before heating. Five-milliliter aliquots of protein dispersions were added to a capped glass vial (10 mL). For the temperature test, samples were incubated at 60–90 °C for 30 min; for the test of protein concentration, pH and ionic strength, samples were incubated at 90 °C for 30 min in a water bath. After heating, the vials were immediately transferred to an ice-water bath, in which the heat-induced particle dispersions were maintained at 4 °C.

2.9. Turbidity Measurement

Particle dispersions (1%, w/v) were diluted with distilled water at a ratio of 1:10. The absorbances of sample solutions at 500 nm were measured with a UH5300 spectrophotometer (Hitachi, Tokyo, Japan). Turbidity was expressed by light transmittance.

2.10. Determination of Z-Average Hydrodynamic Diameters and ζ-Potentials

A Zetasizer Nano ZS instrument (Malvern Instruments, Worcestershire, UK) was employed to measure diameters of particles and the ζ-potential [17]. For Z-average hydrodynamic diameter measurement, the protein dispersion/emulsion was diluted with double-distilled water (1:200, v/v). After equilibrating for 120 s, samples were measured with 15 sequential readings. For ζ-potential measurement, the protein dispersions/ emulsions were diluted at a ratio of 1:200 (v/v) using double-distilled water with equivalent pHs. Samples were measured in triplicate, 12 sequential readings per measurement.

2.11. Atomic Force Microscope (AFM)

The surface morphology of particles was measured using AFM [7]. The protein solutions were diluted to a protein concentration of 5 μg/mL. A droplet (1–3 μL) of prepared sample was spread on a freshly cleaved mica disk and air dried naturally at room temperature. The images were obtained using a Dimension ICON microscope (Bruker Corporation, Jena, Germany).

2.12. Emulsifying Activity Indexe (EAI) and Emulsifying Stability Index (ESI)

EAI and ESI were determined by turbidimetric method [18]. Samples were dissolved or diluted with distilled water and adjusted to pH 7.0 with 0.1 M NaOH and HCl to obtain 0.1% (w/v) aqueous solutions. 10 mL of corn oil in 30 mL of sample solution was emulsified by an Ultra-Turrax blender (IKA T25 Basic, Staufen, Germany) at 11,000 rpm for one minute at room temperature. One hundred microlitres of the emulsions were pipetted from the bottom of the tube into 5 mL of SDS solutions (0.1%, w/v) immediately (0 min) and 10 min after homogenization. The turbidity of the diluted solutions was then determined at 500 nm. The EAI was the absorbance taken immediately after emulsification. The ESI values were calculated using the following equations:

$$\text{ESI (min)} = A_0 \times 10/(A_0 - A_{10}) \qquad (2)$$

where A_0 and A_{10} are the absorbances of diluted emulsions at 0 and 10 min, respectively.

2.13. Statistical Analysis

To check reproducibility, tests were carried out in triplicate. Results were expressed as the mean value ± standard deviation (SD) of three independent treatments. Analysis of variance was used to calculate the significance of the samples, and sample means were separated using the Student's paired t-test. Differences were considered significant at $p < 0.05$.

3. Results and Discussion

3.1. Structure Characteristics of IMO-EWP Conjugates

The degree of glycation (DG) of 2.5%, 10% and 50% IMO-EWP were 62.73, 68.17 and 71.05, respectively, which indicated that the higher IMO ratios resulted in higher DG of glycated EWP. Electrophoretic patterns of EWP and its conjugates with IMO were shown in Figure 1. Native EWP was mainly composed of three major bands approximately at 14, 45 and 76 kDa, which correspond to lysozyme (LZ), ovalbumin (OVA) and ovotransferrin (OT), respectively [19]. After the MR, changes happened in the MW as seen by the densities of the bands (lanes 3–5). The bands of LZ, OVA and OT shifted to higher MW, and a large number of continuous bands appeared on the top of the conjugate electrophoretic patterns, indicating the formation of protein polymers in glycated EWP. The results were consistent with previous studies [13,20]. The MW (Figure 1a) and saccharide moieties (Figure 1b) of LZ, OVA and OT gradually increased with the increase of DG, which further confirmed the formation of conjugates (Table 1).

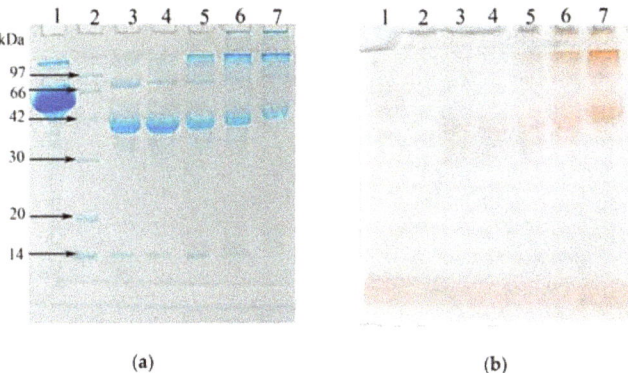

Figure 1. SDS-PAGE profiles of egg white protein (EWP) and isomalto-oligosaccharide (IMO)-EWP conjugates stained for proteins (**a**) or saccharides (**b**). Lane 1: Bovine albumin (negative control); Lane 2: marker proteins; Lane 3: EWP; Lanes 4: EWP incubated at glycation condition for 3 days; Lane 5–7: IMO-EWP conjugates with sugar-protein weight ratios of 2.5%, 10% and 50%, respectively.

FTIR was employed to analyze the secondary structure of EWP and 50% IMO-EWP conjugate (Figure 2). The hydroxyl stretching bands (~3300 cm^{-1}) of IMO-EWP sample showed stronger absorption, indicating that the conjugation of IMO could increase the hydroxyl content of EWP. Meanwhile, the absorption peaks of EWP at 1652 and 1538 cm^{-1} shifted to 1664 and 1547 cm^{-1} after glycation. A previous study reported that the amide I absorption band at 1664 cm^{-1} represent the structure formation of Schiff bases with spectral overlap of the C=O group, coupled with in-plane –NH bending and C=N linkage [21], and the absorption band at 1538 cm^{-1} was attributed to the primary amino group [22]. The above results suggested that the condensation reaction of primary amino groups of protein with carbonyl groups of reducing sugar was happened to form Schiff base products with the release of water and consumption of amino groups.

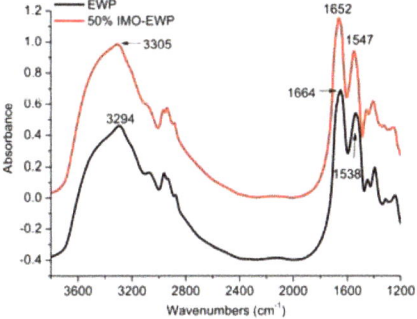

Figure 2. Fourier transform infrared spectroscopy (FTIR) spectra of EWP and 50% IMO-EWP conjugate at pH 7.0.

3.2. Thermal Denaturation Properties of IMO-EWP Conjugates

The DSC curves of EWP and IMO-EWP conjugates during programmed heating at pH 7.0 are shown in Figure 3. The EWP exhibited two major endothermic transitions: the first peak at 66.56 ± 0.03 °C, which mainly raised from the denaturation of OT, and the second peak at 84.11 ± 0.05 °C indicated the denaturation of OVA. These results are similar to previous studies [23]. Notably, the endothermic peak of OT in IMO-EWP conjugates disappeared and the endothermic peak of OVA broadened with an increase of IMO ratio during glycation. It is reported that the broadening of the peak indicates the existence of denatured intermediates different from the native form [24]. This intermediate state, referred to as 'molten globule' state [25], maintained a native like secondary structure but tends to lose some of its tertiary structure. Because the temperature of MR was near the denaturation temperatures (T_d) of OT, the OT turned to 'molten globule' during glycation. In addition, glycation partially denatured the OVA to 'molten globule' state and increasing IMO ratios could significantly promote the conformation changes. The DSC characteristic changes for heat denaturation of EWP and IMO-EWP conjugates are summarized in Table 1. The T_d of OVA in the conjugates was significantly higher than that of EWP, and T_d which elevated with an increase in IMO ratios. Generally, for a globular protein, a higher T_d is related to higher thermal stability [11]. Thus, the thermal stability of EWP was remarkably improved by glycation, in agreement with previous studies [11,13]. Conversely, the enthalpy changes (ΔH) of OVA in the conjugates were lower, as compared to EWP. With an increase of IMO ratio, the ΔH value of OVA gradually decreased. The lower ΔH values of OVA in the conjugates were attributed to the partial unfolding of OVA during glycation [11]. In conclusion, glycation could change the conformation of EWP and significantly improve its thermal stability.

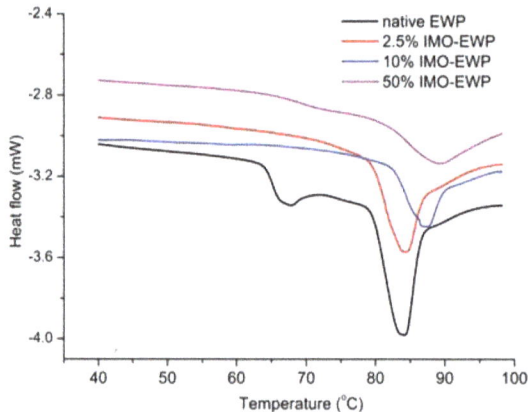

Figure 3. Differential Scanning Calorimetry (DSC) thermograms of EWP and IMO-EWP conjugates at pH 7.0.

Table 1. DSC assessment of denaturation temperature (T_d) and the associated enthalpy change (ΔH) of EWP and IMO-EWP conjugates at pH 7.0.

Samples	Peak 1		Peak 2	
	T_d (°C)	ΔH (J/g)	T_d (°C)	ΔH (J/g)
EWP	66.56 ± 0.04	0.67 ± 0.03	84.11 ± 0.05 [a]	3.94 ± 0.02 [a]
2.5% IMO-EWP	-	-	84.06 ± 0.12 [a]	3.07 ± 0.04 [b]
10% IMO-EWP	-	-	87.19 ± 0.02 [b]	2.84 ± 0.05 [c]
50% IMO-EWP	-	-	89.17 ± 0.06 [c]	2.74 ± 0.02 [c]

Different letters (a–c) in the same column indicate significant differences ($p < 0.05$).

3.3. Surface Hydrophobicity (H_o) and ζ-Potential of IMO-EWP Conjugates

Hydrophobic interaction is a major attractive intermolecular force to facilitate protein aggregation, and therefore, the H_o of proteins was measured using the ANS method. The H_o of 2.5%, 10%

and 50% IMO-EWP were 8742 ± 236, 6074 ± 101 and 3102 ± 1391, respectively, while that of EWP was 1313 ± 955. The increased H_o of the conjugates was much higher than that of the EWP. Previous studies suggest that denaturation of OVA during glycation would unfold the native tertiary structure and expose the buried hydrophobic groups [13,24]. The changes in structure of EWP may relate to the decrease of ΔH values and the improvement of thermal stability after glycation (Table 1). In addition, the H_o is highest for 2.5% IMO-EWP and gradually decreased with the increase of DG, which was due to that the increased attachment of sugar chains decreasing the hydrophobicity of proteins [26].

As shown in Figure 4, the ζ-potentials of conjugates were consistently lower compared to EWP. When pH was above pI, the absolute values of ζ-potentials for the conjugates were much higher than EWP, which means more electrostatic repulsion among the IMO-EWP molecules. It has been reported that the reducing end carbonyl groups of saccharides were mainly attached to the lysine and arginine residues (positively charged) of proteins during glycation, which would reduce the positive electrical charges [27]. On the other hand, more charged amino acids would be exposed to the exterior due to the protein unfolding caused by the dry-heating process of glycation.

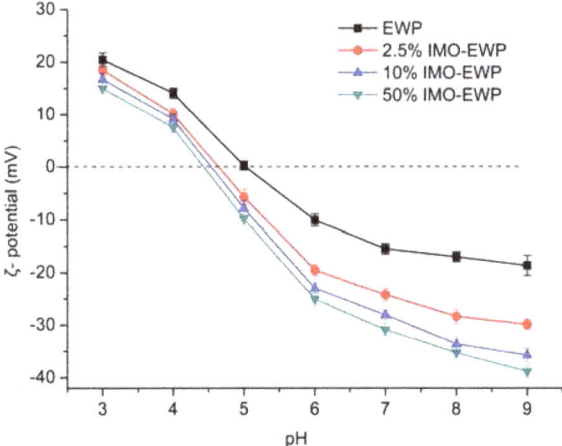

Figure 4. ζ-potentials of EWP and IMO-EWP conjugates at pH 3.0–9.0.

3.4. Effects of Temperature, Protein Concentration, pH and Ionic Strength on the Formation of Nanoparticles

Heating temperature has a significant influence on particle size. Changes in turbidity could reflect the sizes of the heat-induced particles. As shown in Figure 5a, the turbidity of the heat-induced particles formed from the EWP were significantly increased when the heating temperatures were above 60 °C, whereas the heat-induced particles formed from the IMO-EWP conjugates scarcely changed their turbidity and remained transparent with the increase of heating temperatures. Because the hydrophobic interactions were strengthened after glycation (Table 1), the inhibition of conjugate aggregation after heating may be attributed to the increased thermal stability of OT (Figure 3) and electrostatic repulsion (Figure 4). Sponton et al. also found that the electrostatic repulsions among EWP anionic charges at extremely alkaline pH were the dominating mechanism responsible for the suppression of thermal aggregation [8].

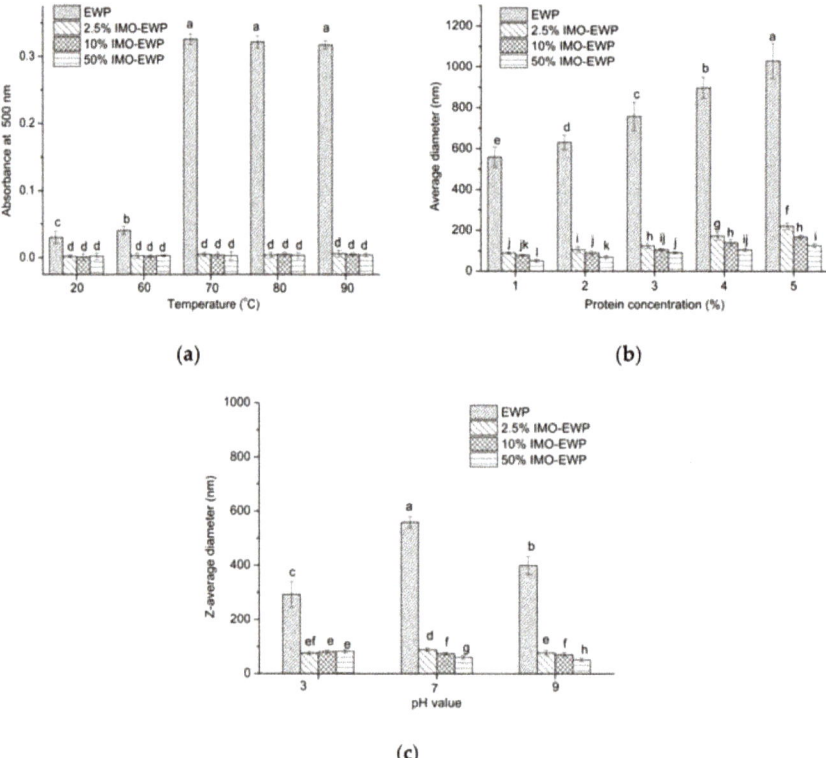

Figure 5. Turbidity of the heat-induced particles formed from EWP and IMO-EWP conjugates at different temperatures (a). Z-average hydrodynamic diameters of heat-induced particles formed at different protein concentrations (b) and pHs (c) from EWPs and conjugates. Different letters (a–l) indicate significant differences ($p < 0.05$).

As shown in Figure 5b, the particle size of aggregates increased with protein concentration. This could be explained by a rise in the number of protein macromolecules increasing the molecular collision, promoting heat-induced aggregation [28]. Besides, the diameters of the particles produced from conjugates were <200 nm, and considered "nanoparticles". The effects of pH and ionic strength on the formation of heat-induced particles were also considered (Figures 5 and 6). The particle dispersions of the conjugates were yellowish and the colors were deepened with increased of DG, which was a characteristic of nonenzymatic browning due to MR. It can be seen that the overall particle dispersions showed turbid appearance (Figure 6) and large particle size at pH 5.0 (the sizes of EWP aggregates at pH 5.0 were too large to be determined by DLS), which would indicate the presence of protein aggregates [1]. This was a result of the electrostatic repulsion being weakened as pH approaches pI [29]. Without the addition of NaCl, the heat-induced particles formed from the IMO-EWP conjugates maintained their clarity at pH 3.0, 7.0, and 9.0, while the particles from the EWP samples were transparent only at pH 3.0. This was consistent with the diameter results (Figure 5c). The increased electrostatic repulsion induced by glycation was responsible for the transparency of nanoparticles at pH 7.0–9.0. However, the transparent particle dispersions became turbid when NaCl was added, and the turbidity became more severe with the increased addition of NaCl. This was due to the reduction in electrostatic repulsion among the protein molecules caused by binding or shielding of NaCl [30]. Besides, the particles formed from the conjugates with higher DG showed more clarity appearance at the same pH and NaCl conditions (Figure 7c,d). These results again agreed with the decreased particle size with the increase of DG (Figure 5c). In conclusion, the heat-induced particles tended to lower sizes with the increase of DG, pH and ionic strength at lower protein concentrations.

Figure 6. Photographs of EWP and IMO-EWP heat-induced particles prepared at different pHs and NaCl concentrations. EWP (**a**), 2.5% IMO-EWP (**b**), 10% IMO-EWP (**c**) and 50% IMO-EWP (**d**). Samples were adjusted to pH 3.0, 5.0, 7.0, 9.0 (vials from left to right in each image) and 0–150 mM NaCl was added as appropriate before heating.

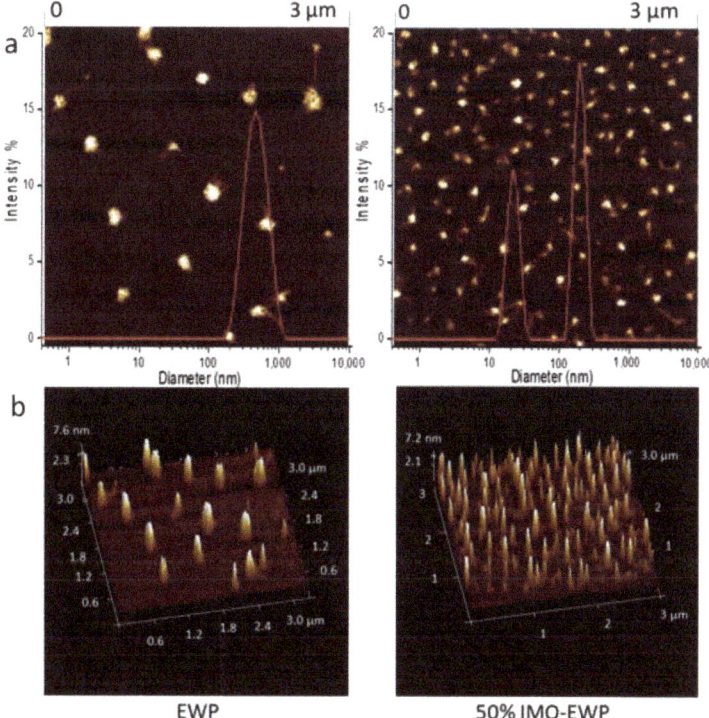

Figure 7. Atomic force microscope (AFM) images and particle diameter distributions of heat-induced particles from EWP and 50% IMO-EWP conjugate. 2D images and particle diameter distributions (**a**); 3D images (**b**).

To confirm the micromorphology properties, the heat-induced particles fabricated by EWP and conjugate at pH 7.0 were observed under AFM (Figure 7). The dimension of the conjugate particle was much smaller than EWP particle, which agreed with the hydrodynamic diameter determined by DLS (Figure 5).

3.5. Emulsifying Activity Index (EAI) and Emulsifying Stability Index (ESI) of Nanoparticles

Proteins can be adsorbed at the oil-water interface to form a coherent viscoelastic layer to stabilize the oil. As shown in Figure 8, EAI and ESI of the conjugates (B–D) and their nanoparticles (F–H) were remarkably higher than EWP (A). Moreover, the EAI and ESI of the nanoparticles from conjugates were increased about 3-folds and 2-folds, respectively, as compared with EWP. This can be explained by the increased H_o induced by MR and heat treatment [5] which could promote the adsorption of protein on oil-water interface and shielded the oil droplets of emulsions against aggregation [31]. This agrees with previous reports that the emulsifying properties of EWP or its constituent proteins were improved through MR or heat treatment [13,32]. In addition, the ESI of conjugates and their nanoparticles were enhanced with the increase of DG, which was likely related to the increased electrostatic repulsion (Figure 4). In conclusion, the heat-induced nanoparticles formed from EWP conjugates showed the best emulsifying activity and stability.

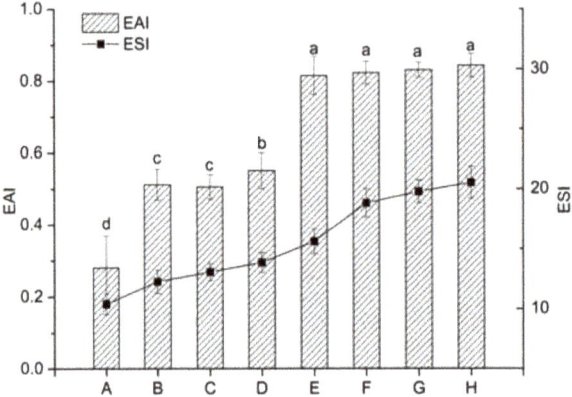

Figure 8. Emulsifying activity and stability of the unheated and heat-induced EWPs at pH 7.0. A–D: unheated EWP, 2.5% IMO-EWP, 10% IMO-EWP and 50% IMO-EWP, respectively; E–H: heat-induced EWP, 2.5% IMO-EWP, 10% IMO-EWP and 50% IMO-EWP, respectively. Different letters (a–d) on the column indicate significant differences ($p < 0.05$).

4. Conclusions

Glycation of EWP with IMO changed the conformation of EWP and improved its thermal stability and surface hydrophobicity. Moreover, the increased absolute values of ζ-potential after glycation would enhance the electrostatic repulsion among the IMO-EWP molecules when pH was above pI. As a result, thermal aggregation of IMO-EWP conjugates was significantly suppressed and transparent nanoparticle dispersions (with particle diameter <200 nm) were obtained after heating at pH 7.0–9.0. This demonstrated that the heat-induced nanoparticles showed the highest emulsifying activity and stability. The present study made a successful attempt in fabrication of heat-induced nanoparticles with improved emulsifying activity and stability at mild pH and salt conditions. It provided the supplement for EWP nanoparticle application on neutral conditions, compared to the aggregates fabricated at extremely acid and alkaline conditions. Further investigation on the storage stability and encapsulation of bioactive compounds using nanoparticles will expand its application in many food systems, such as EWP beverages or stabilizers for emulsions.

Supplementary Materials: The following are available online at http://www.mdpi.com/2079-4991/8/11/943/s1, Figure S1: High performance liquid chromatography (HPLC) of IMO, Table S1: Composition of IMO determined by HPLC (Figure S1).

Author Contributions: C.W. and Y.Y. conceived and designed the experiments; C.W. performed the experiments and analyzed the data; C.W. wrote the original paper; X.R. and Y.S. edited the manuscript.

Funding: This research was funded by the National Key Research and Development Program of China, grant number 2018YFD0400303; the National Natural Science Foundation of China, grant number No. 31671809 and Jiangsu province "Collaborative Innovation Center for Food Safety and Quality Control" industry development program.

Acknowledgments: The authors thank Fang Qin, Yun Ma, Chen Chen for providing valuable guidance for the use of equipment.

Conflicts of Interest: The authors declare no conflict of interest.

References

1. Liu, Y.; Oey, I.; Bremer, P.; Carne, A.; Silcock, P. Effects of pH, temperature and pulsed electric fields on the turbidity and protein aggregation of ovomucin-depleted egg white. *Food Res. Int.* **2017**, *91*, 161–170. [CrossRef] [PubMed]
2. Nicolai, T. Formation and functionality of self-assembled whey protein microgels. *Colloids Surf. B Biointerfaces* **2016**, *137*, 32–38. [CrossRef] [PubMed]
3. Yildiz, G.; Andrade, J.; Engeseth, N.E.; Feng, H. Functionalizing soy protein nano-aggregates with pH-shifting and mano-thermo-sonication. *J. Colloid Interface Sci.* **2017**, *505*, 836–846. [CrossRef] [PubMed]
4. Vries, A.D.; Wesseling, A.; Linden, E.V.D.; Scholten, E. Protein oleogels from heat-set whey protein aggregates. *J. Colloid Interface Sci.* **2017**, *486*, 75–83. [CrossRef] [PubMed]
5. Croguennec, T.; Renault, A.; Beaufils, S.; Dubois, J.-J.; Pezennec, S. Interfacial properties of heat-treated ovalbumin. *J. Colloid Interface Sci.* **2007**, *315*, 627–636. [CrossRef] [PubMed]
6. Campbell, L.; Raikos, V.; Euston, S.R. Modification of functional properties of egg-white proteins. *Nahrung/Food* **2003**, *47*, 369–376. [CrossRef] [PubMed]
7. Chang, C.; Niu, F.; Gu, L.; Li, X.; Yang, H.; Zhou, B.; Wang, J.; Su, Y.; Yang, Y. Formation of fibrous or granular egg white protein microparticles and properties of the integrated emulsions. *Food Hydrocoll.* **2016**, *61*, 477–486. [CrossRef]
8. Sponton, O.E.; Perez, A.A.; Ramel, J.V.; Santiago, L.G. Protein nanovehicles produced from egg white. Part 1: Effect of pH and heat treatment time on particle size and binding capacity. *Food Hydrocoll.* **2017**, *73*, 67–73. [CrossRef]
9. Liu, G.; Zhong, Q. Thermal aggregation properties of whey protein glycated with various saccharides. *Food Hydrocoll.* **2013**, *32*, 87–96. [CrossRef]
10. Xu, C.H.; Yang, X.Q.; Yu, S.J.; Qi, J.R.; Guo, R.; Sun, W.W.; Yao, Y.J.; Zhao, M. The effect of glycosylation with dextran chains of differing lengths on the thermal aggregation of β-conglycinin and glycinin. *Food Res. Int.* **2010**, *43*, 2270–2276. [CrossRef]
11. Liu, Y.; Zhao, G.; Zhao, M.; Ren, J.; Yang, B. Improvement of functional properties of peanut protein isolate by conjugation with dextran through Maillard reaction. *Food Chem.* **2012**, *131*, 901–906. [CrossRef]
12. Cai, L.; Lin, C.; Yang, N.; Huang, Z.; Miao, S.; Chen, X.; Pan, J.; Rao, P.; Liu, S. Preparation and characterization of nanoparticles made from co-incubation of SOD and glucose. *Nanomaterials* **2017**, *7*, 458. [CrossRef] [PubMed]
13. An, Y.; Cui, B.; Wang, Y.; Jin, W.; Geng, X.; Yan, X.; Li, B. Functional properties of ovalbumin glycosylated with carboxymethyl cellulose of different substitution degree. *Food Hydrocoll.* **2014**, *40*, 1–8. [CrossRef]
14. Rao, Q.; Rocca-Smith, J.R.; Schoenfuss, T.C.; Labuza, T.P. Accelerated shelf-life testing of quality loss for a commercial hydrolysed hen egg white powder. *Food Chem.* **2012**, *135*, 464–472. [CrossRef] [PubMed]
15. Byler, D.M.; Susi, H. Examination of the secondary structure of proteins by deconvolved FTIR spectra. *Biopolymers* **1986**, *25*, 469–487. [CrossRef] [PubMed]
16. Hayakawa, S.; Nakai, S. Relationships of hydrophobicity and net charge to the solubility of milk and soy proteins. *J. Food Sci.* **1985**, *50*, 486–491. [CrossRef]
17. Chang, C.; Niu, F.; Su, Y.S.; Qiu, Y.; Gu, L.; Yang, Y. Characteristics and emulsifying properties of acid and acid-heat induced egg white protein. *Food Hydrocoll.* **2016**, *54*, 342–350. [CrossRef]
18. Jing, H.; Yap, M.; Wong, P.Y.Y.; Kitts, D.D. Comparison of physicochemical and antioxidant properties of egg-white proteins and fructose and inulin Maillard reaction products. *Food Bioprocess Technol.* **2011**, *4*, 1489–1496. [CrossRef]

19. Kato, A.; Minaki, K.; Kobayashi, K. Improvement of emulsifying properties of egg white proteins by the attachment of polysaccharide through Maillard reaction in a dry state. *J. Agric. Food Chem.* **1993**, *41*, 540–543. [CrossRef]
20. Sun, W.W.; Yu, S.J.; Zeng, X.A.; Yang, X.Q.; Jia, X. Properties of whey protein isolate-dextran conjugate prepared using pulsed electric field. *Food Res. Int.* **2011**, *44*, 1052–1058. [CrossRef]
21. Kosaraju, S.L.; Weerakkody, R.; Augustin, M.A. Chitosan-glucose conjugates: Influence of extent of Maillard reaction on antioxidant properties. *J. Agric. Food Chem.* **2010**, *58*, 12449–12455. [CrossRef] [PubMed]
22. Umemura, K.; Kawai, S. Preparation and characterization of Maillard reacted chitosan films with hemicellulose model compounds. *J. Appl. Polym. Sci.* **2008**, *108*, 2481–2487. [CrossRef]
23. Nafchi, A.M.; Tabatabaei, R.H.; Pashania, B.; Rajabi, H.Z.; Karim, A.A. Effects of ascorbic acid and sugars on solubility, thermal, and mechanical properties of egg white protein gels. *Int. J. Biol. Macromol.* **2013**, *62*, 397–404. [CrossRef] [PubMed]
24. Myers, C.D. Study of thermodynamics and kinetics of protein stability by thermal analysis. In *Thermal Analysis of Foods*; Harwalker, V.R., Ma, C.Y., Eds.; Elsevier Applied Science: New York, USA, 1990; pp. 16–50.
25. Jiang, J.; Xiong, Y.L.; Chen, J. pH shifting alters solubility characteristics and thermal stability of soy protein isolate and its globulin fractions in different pH, salt concentration, and temperature conditions. *J. Agric. Food Chem.* **2010**, *58*, 8035–8042. [CrossRef] [PubMed]
26. Broersen, K.; Elshof, M.; De Groot, J.; Voragen, A.G.; Hamer, R.J.; De Jongh, H.H. Aggregation of β-lactoglobulin regulated by glucosylation. *J. Agric. Food Chem.* **2007**, *55*, 2431–2437. [CrossRef] [PubMed]
27. Achouri, A.; Boye, J.I.; Yaylayan, V.A.; Yeboah, F.K. Functional properties of glycated soy 11S glycinin. *J. Food Sci.* **2005**, *70*, C269–C274. [CrossRef]
28. Sponton, O.E.; Perez, A.A.; Ramel, J.V.; Santiago, L.G. Protein nanovehicles produced from egg white. Part 2: Effect of protein concentration and spray drying on particle size and linoleic acid binding capacity. *Food Hydrocoll.* **2017**, *77*, 863–869. [CrossRef]
29. Datta, D.; Bhattacharjee, S.; Nath, A.; Das, R.; Bhattacharjee, C.; Datta, S. Separation of ovalbumin from chicken egg white using two-stage ultrafiltration technique. *Sep. Purif. Technol.* **2009**, *66*, 353–361. [CrossRef]
30. Hegg, P.O.; Martens, H.; Löfqvist, B. Effects of pH and neutral salts on the formation and quality of thermal aggregates of ovalbumin. A study on thermal aggregation and denaturation. *J. Sci. Food Agric.* **1979**, *30*, 981–993. [CrossRef]
31. Hiller, B.; Lorenzen, P.C. Functional properties of milk proteins as affected by Maillard reaction induced oligomerisation. *Food Res. Int.* **2010**, *43*, 1155–1166. [CrossRef]
32. Medrano, A.; Abirached, C.; Moyna, P.; Panizzolo, L.; Añón, M.C. The effect of glycation on oilewater emulsion properties of β-lactoglobulin. *LWT-Food Sci. Technol.* **2012**, *45*, 253–260. [CrossRef]

© 2018 by the authors. Licensee MDPI, Basel, Switzerland. This article is an open access article distributed under the terms and conditions of the Creative Commons Attribution (CC BY) license (http://creativecommons.org/licenses/by/4.0/).

Article

A Novel Fast Photothermal Therapy Using Hot Spots of Gold Nanorods for Malignant Melanoma Cells

Yanhua Yao [1], Nannan Zhang [1], Xiao Liu [1], Qiaofeng Dai [1], Haiying Liu [1], Zhongchao Wei [1], Shaolong Tie [2], Yinyin Li [3], Haihua Fan [1,*] and Sheng Lan [1,*]

1. Guangdong Provincial Key Laboratory of Nanophotonic Functional Materials and Devices, School of Information and Optoelectronic Science and Engineering, South China Normal University, Guangzhou 510006, China; m15625107214@163.com (Y.Y.); clzhangnannan@163.com (N.Z.); shawer1004@163.com (X.L.); daiqf@scnu.edu.cn (Q.D.); hyliu@vip.163.com (H.L.); weizhongchao@263.net (Z.W.)
2. School of Chemistry and Environment, South China Normal University, Guangzhou 510006, China; ties12008@163.com
3. School of Life Sciences, Sun Yat-Sen University, State Key Lab for Biocontrol, Guangzhou 510275, China; liyinyin@mail.sysu.edu.cn
* Correspondence: 20111249@m.scnu.edu.cn (H.F.); slan@scnu.edu.cn (S.L.)

Received: 18 September 2018; Accepted: 25 October 2018; Published: 28 October 2018

Abstract: In this paper, the plasmon resonance effects of gold nanorods was used to achieve rapid photothermal therapy for malignant melanoma cells (A375 cells). After incubation with A375 cells for 24 h, gold nanorods were taken up by the cells and gold nanorod clusters were formed naturally in the organelles of A375 cells. After analyzing the angle and space between the nanorods in clusters, a series of numerical simulations were performed and the results show that the plasmon resonance coupling between the gold nanorods can lead to a field enhancement of up to 60 times. Such high energy localization causes the temperature around the nanorods to rise rapidly and induce cell death. In this treatment, a laser as low as 9.3 mW was used to irradiate a single cell for 20 s and the cell died two h later. The cell death time can also be controlled by changing the power of laser which is focused on the cells. The advantage of this therapy is low laser treatment power, short treatment time, and small treatment range. As a result, the damage of the normal tissue by the photothermal effect can be greatly avoided.

Keywords: gold nanorods; A375 cells; plasmonic coupling; photothermal therapy; hot spot

1. Introduction

Gold nanorods (GNRs) are increasingly receiving academic attention due to their special, unique chemical and optical properties in addition to their high biocompatibility [1–6]. The size of the GNRs is easily controlled, and by adjusting the size of the GNRs, the absorption peaks are modulated in the near-infrared (NIR) region. Nano-sized particles are also easily taken up by cells. Therefore, GNRs show more and more application potential in photothermal therapy. The longitudinal localized surface plasmon resonance in the infrared region of GNRs renders them good materials for cancer photothermal therapy under NIR excitation. For optics applied in biology, NIR excitation first offers low scattering and energy absorption, secondly offering maximum irradiation penetration into deeper tissue. Moreover, the autofluorescence emitted from nontargeted tissue can be partially or totally inhibited under NIR excitation. The stability and biocompatibility of GNRs also make them suitable be used in biological medicine. The conduction electrons of GNRs can be excited by coherent light to induce surface plasmon oscillations, which can be used for photothermal therapy [7–12].

The rapid global economic development has caused major environmental degradation and the destruction of ozone in the atmosphere. More UV penetration increases the incidents of skin cancer. As the largest organ of the human body, the skin is a physical barrier against various infections and environments. Skin cancer is a type of cancer that accounts for almost 40% of the world's cancer population [13]. Melanoma is one of the most common types of skin cancer. Melanoma (cancer caused in melanocytes, which are the pigment-containing cells) has a highly tendency to metastasize to other organs of the body. Plasmonic photothermal therapy for the treatment of cancer has received a great deal of attention in recent years. Specifically, in the past decade, there has been much progress in the development of GNRs for photothermal therapy applications due to their localized surface plasmon resonance [14–16] as well as their inherently low toxicities [17–19]. Compared to other cancers, melanoma is mostly located on the surface of the body, making it easy to treat directly with a laser. However, reports of plasmon resonance photothermal therapy for skin cancer cells are lacking. For photothermal therapy, it can be seen that many works focus on the design and modification of photothermal materials. These materials have better properties (such as cell targeting, photothermal conversion efficiency, etc.) and can improve the efficiency of photothermal therapy [20,21]. Some works are focused the effects of photothermal therapy on cells [22]. These studies have led to the further development of photothermal therapy. After summarizing the previous research work, it was found that the range of the traditional photothermal therapy laser action reaches the 2-cm level, the time is generally several minutes to 10 min, and the light source is generally provided by a continuous-wave diode laser. However, there is a lack of research on low-power photothermal therapy on a small scale and in a short period of time. For clinical applications, the energy of the input laser should be as low as possible, the laser treatment range should be as small as possible, and the laser and the biological action time should be as short as possible to avoid damage to healthy tissues. Research on photothermal therapy has been trying to achieving these goals.

In this study, we investigated a low-power rapid photothermal therapy for individual cancer cells by using the plasmonic coupling between GNRs. The cell viability of human malignant melanoma cells (A375) cells still exceeded 80% after incubation with GNRs for 24 h by controlling the concentration of GNRs in the culture medium. This result indicates that the cytotoxicity of GNRs is negligible in the doses used in this work. The transmission electron microscopy (TEM) image of A375 cells shows that the GNR clusters form naturally in the organelles of cells. The simulation results show that the plasma coupling between GNRs can effectively enhance the electromagnetic field in interparticle gap space, known as the hot spot. Therefore, a low-power femtosecond laser can result in rapid acute damage to individual cancer cells by using the hot spot. Necrosis can be induced in A75 cells immediately after they are irradiated by a femtosecond laser as low as 14.6 mW in 20 s. The cell's death time can also be controlled by changing the power of the laser which is focused on the cells. In this therapy, the laser can directly illuminate the skin cancer cells. The power of the irradiation is very low, the irradiation time is very short, the treatment is small in scope, and it can be controlled to treat only individual cells. As a result, the damage of the normal tissue by the photothermal effect can be greatly avoided. This work represents a big step forward in the efforts to employ the plasma photothermal effect in the actual treatment of disease.

2. Materials and Methods

2.1. Synthesis of PEG-Coated GNRs

In this work, the modified seedless method, which is described in detail in Reference [23], was used to synthesize GNRs. The as-prepared GNRs were then stabilized by polyethylene glycol (PEG) [24].

2.2. UV-Vis-NIR Absorption Spectroscopy

The absorption spectra of the GNRs were obtained using an ECAN microplate reader (Tecan, Durham, CA, USA). The solution was put into a quartz cuvette. The absorption spectrum of deionized water or PEG was detected, to be used as a baseline to calibrate the spectra of all the samples.

2.3. Transmission Electron Microscopy

The morphology of the GNRs was examined using transmission electron microscopy (TEM). The TEM images were taken by using a high-resolution TEM (JEOL-JEM-2100HR, JEOL Company, Akishima, Tokyo, Japan) operating at an accelerating voltage of 200 kV. The samples were prepared on 200-mesh copper grids.

2.4. Cell Culture and Cell Viability

The cancer cells used in the photothermal therapy were human malignant melanoma (A375). They were obtained from the Cell Lab of the Cell Resource Center of the Chinese Academy of Sciences. The cell culture methods are very similar to those described in Reference [25].

Cell viability was measured using the 3-(4,5-dimethylthiazol-2-yl)-2,5-diphenyltetrazolium bromide (MTT) assay. This method is similar to those described in Reference 25. The 96-well plate was then put into an iMark Microplate (BioRad) to measure optical densities (OD) at 490 nm. The cell morphology of A375 cells incubated with GNRs at different concentrations are shown in Figure S2.

2.5. Cellular Uptake of Gold Nanorods

The concentration of Au was measured by inductively coupled plasma-mass spectrometry (ICP-MS) (ICAP-qc, Thermo Fisher, Boschstr, Kleve, Germany) [26]]. The number of gold nanorods in a single cell can be obtained by dividing the ICP-MS result by the cell density in the ICP-MS sample and then dividing by the mass of a single gold nanorod. The cellular uptake images were examined using TEM observation. After incubation with the GNRs for 24 h in a humidified incubator (37 °C, 5% CO_2), the cells were washed three times with Phosphate Buffered Solution (PBS) and pelleted by centrifugation. Finally, they were fixed in glutaraldehyde (2.5%), embedded in resin, cut to ultra-thin sections, stained by osmic acids, and finally imaged using TEM at an acceleration voltage of 120 kV. The samples were prepared on Ni mesh with a carbon support film.

2.6. Photothermal Therapy Experiments

The schematic diagram of the experimental setup is shown in Figure S3. In the photothermal therapy experiment, a Ti:sapphire oscillator (Mira 900 S, Coherent, Santa Clara, CA, USA) with a repetition rate of 76 MHz was reflected into an inverted microscope (Axio Observer A1, Zeiss, Santa Clara, CA, USA) and was focused on the targeted cell using a 60× objective lens. The diameter of the laser spot was estimated to be ~1 µm. The photothermal therapy experiment was also carried out using a confocal laser scanning microscope (TCS-SP5, Leica Oberkochen, Germany,). For the photothermal therapy experiment, the cells were incubated with the culture medium containing 78 pM GNRs for 24 h. After that, the culture medium containing GNRs was removed and the new culture medium containing no GNRs was added into the dish. The laser power used for the photothermal therapy ranged from 0 to 30 mW.

2.7. Numerical Simulations

The finite-difference time-domain (FDTD) software developed by Lumerical Solutions, Inc. (Thurlow Street, Vancouver, BC, Canada, http://www.lumerical.com) was employed to simulate the distribution of electric field of GNRs [27,28]. In the FDTD simulations, non-uniform grids with the smallest grid of 0.2 nm and perfectly matched layer conditions were employed.

3. Results and Discussion

3.1. Characteristic of GNRs

It is known from References [29–31] that plasmon resonance coupling between gold nanorods can generate large amounts of heat. In order to apply the plasmon resonance characteristics of gold nanorods to cell photothermal therapy, the optical properties and cytotoxicity of gold nanorods were studied. The modified seedless method described in Reference [21] was used to synthesize the GNRs employed in this paper. The short diameter of the GNRs was 7 ± 1.3 nm and the long diameter was 27 ± 6.1 nm. The aspect ratio (ratio of the long diameter to the short diameter) was 3.75 ± 0.73. Figure 1a shows the transmission electron microscopy (TEM) images of the GNRs. It can be seen from Figure 1a that the GNRs are regular in shape and uniform in size. Figure S1 show the GNRs average aspect ratio is 3.75, the standard deviation is 0.73. Figure 1b depicts the absorption spectra of the GNRs. The absorption peaks show that the longitudinal surface plasmon resonance (LSPR) of GNRs was located at 800 nm. Based on the plasmon resonance peak with GNRs at 800 nm, an 800-nm laser was used as excitation light in the photothermal experiment below. The cytotoxicity of GNRs to A375 cells was examined by MTT assay. In this case, the A375 cells were incubated with the GNRs of different concentrations ranging from 26 to 130 pM. The highest cell viability was 98%, corresponding to the concentration of GNRs at 26 pM; the lowest cell viability was 81%, corresponding to the concentration of GNRs at 130 pM. From Figure 1c, it can be seen that the GNRs can be considered to be nontoxic to A375 cells, as the viability of the A375 cells still exceeds 80% at the highest concentration of the GNRs after being incubated with the GNRs for 24 h. The uptake of the GNRs for A375 cells was also studied in order to realize efficient photothermal therapy. As shown in Figure 1d, it was found that the uptake of the GNRs and the concentration of the GNRs do not exhibit a linear relationship. When the culture concentration of GNRs was 78 pM, the cellular uptake of GNRs was 1360, which was the highest among the numerous culture concentrations. In order to achieve effective photothermal therapy, a culture medium with a concentration of 76 pM GNRs was used in the following experiments, because of the high viability and high uptake at this concentration.

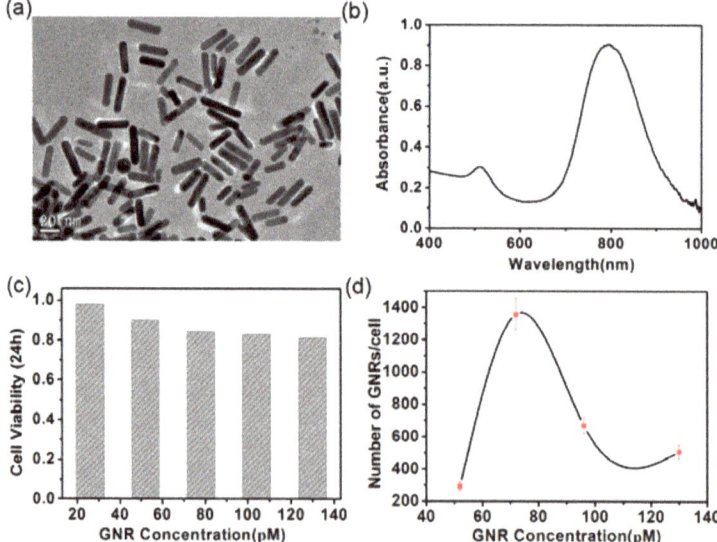

Figure 1. (a) TEM image of the synthesized gold nanorods (GNRs); (b) absorption spectrum of the GNRs dispersed in water; (c) cytotoxicity of GNRs against A375 cells; (d) uptake of the GNRs measured for A375 cells. Error bars represent the standard deviation of three experiments.

3.2. Interaction between the GNRs and A375 Cells

In order to observe the cellular uptake and localization of GNRs in A375 cells, the TEM measurements of cells were preformed after incubating the cells with GNRs for 24 h. As shown in Figure 2a, some nanoparticles were found in intracellular vesicular organelles such as lysosomes. Figure 2b–d show the TEM images of the cells incubated with GNRs for 24 h at different resolutions. The energy dispersive X-ray spectroscopy (EDX) results revealed that these nanoparticles in the cells were GNRs (see Figure S4). The other element peaks observed in the EDX spectrum come from the Ni mesh with a carbon support film and from the dyes used in the cell. Combined with information from TEM images and EDX results, it was confirmed that after 24 h of co-culture with A375 cells, GNRs were taken up by cells and accumulated in vesicular organelles after entering into cells. The GNR clusters formed naturally in lysosomes. From the TEM image of the cells, it can be seen that the gold nanorods in the vesicular organelles are at various angles with each other. Some dimers of GNRs are labeled in Figure 2c,d, and the enlarged view of these dimers of GNRs show that the angles between the gold nanorods vary from 0° to 180°.

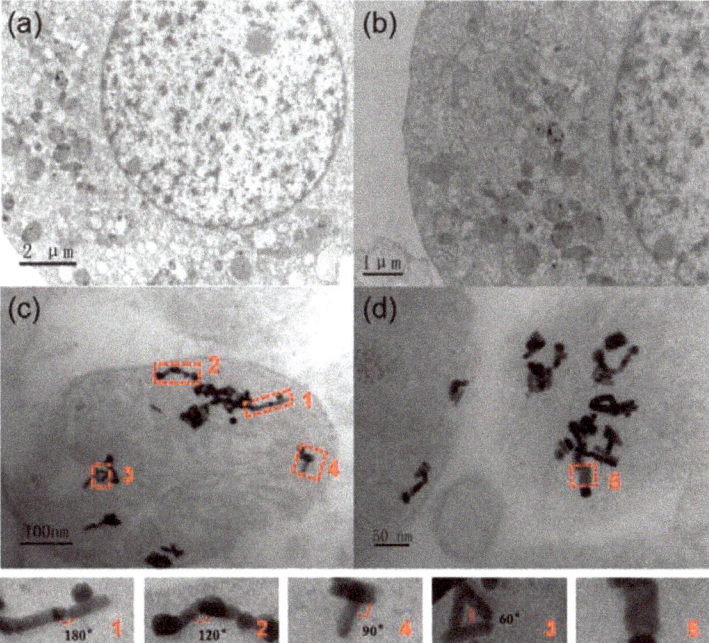

Figure 2. (a) and (b) are TEM images of an A375 cell that has been cultured with GNRs for 24 h (c) and (d) are TEM images of GNRs cluster naturally created in the lysosome of the A375 cell.

3.3. The Field Enhancement of GNRs

If the GNRs are excited by an NIR laser, upon surface plasmon formation, nonradiative relaxation occurs through electron–phonon and phonon–phonon coupling, efficiently generating localized heat that can be transferred to the surrounding environment. The photothermal conversion heat in the experiment may correspond to the field enhancement around the particles. Based on the spacing and angles of the gold nanorods in the organelles labeled in Figure 2, we performed a numerical simulation to clarify the electric field enhancement around the GNRs. Figure 3 shows the near-field intensity map of two gold nanorod systems (LSPR 800 nm) at 800 nm laser excitation. An interparticle separation of 1 nm was considered for the calculation, which is in agreement with the length of the

TEM image. The field enhancement ($|E|/|E_0|$) surrounding the GNRs for both z and x polarization was studied. In the case where the incident light is polarized in the z-direction, when the two gold rods are at a 180-degree angle and the pitch is 1 nm, the field-strength between the gold nanorods is as high as 60 times. This means that the laser intensity $|E|^2$ at this position will increase by 3600 times. If the angle between the two GNRs changes, the field enhancement value will vary from 50 to 58. For the case where the incident light is polarized in the x-direction, when the two gold nanorods are at 0 degrees, the field enhancement will reach a maximum of 5 times, and at 120 degrees, it will be a minimum of about 1.4 times. The minimum field enhancement of 1.4 times corresponds to a laser intensity enhancement of 1.96 times. Thus, no matter which polarization direction of incident light, after interacting with the gold nanorods, it will cause field enhancement due to the plasmon resonance coupling, which will generate heat to increase the temperature of the cells and cause necrosis. It can be seen that if the laser can be aimed at the GNR clusters naturally formed in the organelles, the generated heat can achieve single-cell ultra-low energy photothermal therapy.

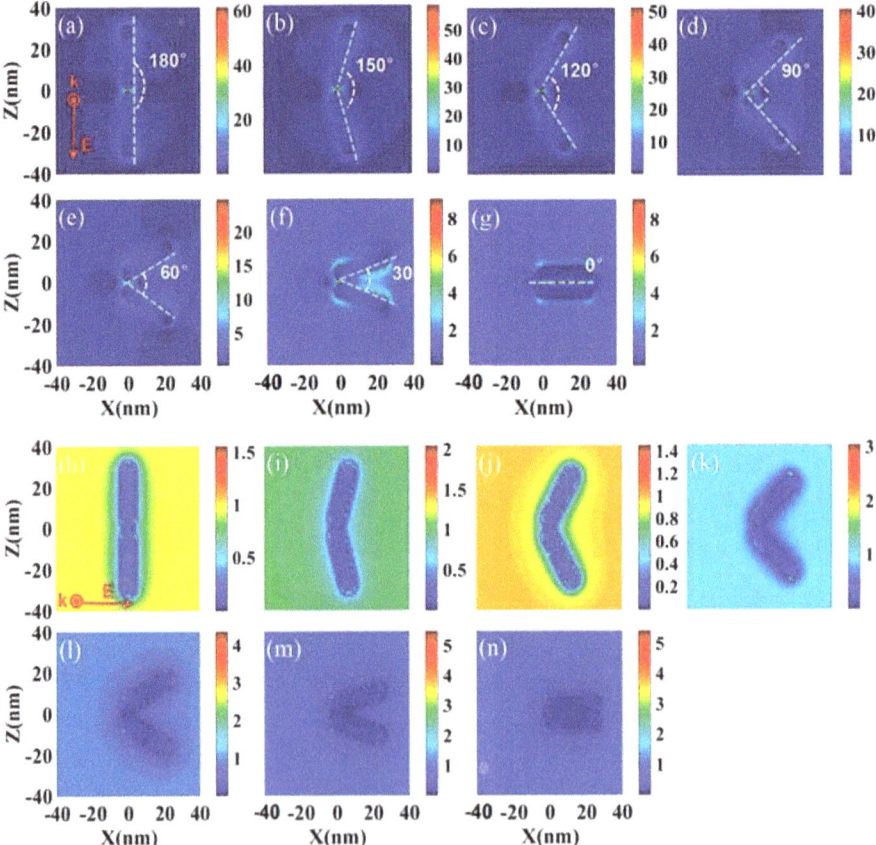

Figure 3. The near-field intensity map of two gold nanorods systems (LSPR 800 nm) at 800 nm laser excitation; the interparticle separation is 1 nm and the angle between the two nanorods varies from 0 degrees to 180 degrees. Both z and x polarization are studied. Images (**a–g**) show the z polarization; images (**h–n**) show the x polarization.

3.4. The Photothermal Therapy of GNRs

After analyzing the numerical simulation results, we conducted a series of experiments to study the interaction between the laser and the cells incubated with the GNRs. After incubating A375 cells with the GNRs for 24 h, the GNRs entered into cells and accumulated in the organelles (such as lysosomes) of the cells and formed GNR clusters. The position of the gold nanorod cluster in the cell could be obtained by scanning the cell with an 800-nm laser and observing the two-photon-absorption induced luminescence (TPL) emission image of the cell. The power of the laser used to scan the cells was below the cell's damage threshold. After determining the location of the GNR cluster, a high-power femtosecond laser could be used to simply excite the GNR clusters. In this way, an accurate and rapid injury to a single A375 cell can be easily achieved.

Figure 4 shows the process of laser treatment to a single A375 cell. Figure 4a is the bright field image of an A375 cell incubated with GNRs for 24 h. This cell was scanned with a laser with a power of 4 mW, and a two-photon emission image of the cell was obtained. Figure 4b shows the combination of the image of the bright field and the TPL emission image. The location of GNR clusters can be found by confirming the TPL emission area (indicated by the arrow in Figure 4b). Then, a high-power laser (30 mW) was use to illuminate the point indicated by the arrow in Figure 5b. The laser does not scan when it hits this point. After the laser lasts for 0.1 s, a 4-mW laser was used to rapidly scan the cell to obtain its morphology and two-photon fluorescence images. This makes it easier to study the series of changes that occur after the laser is applied to the gold nanorod clusters. From Figure 4c–l, the morphological changes of cells under different laser irradiation times are shown. From Figure 4, it can be seen that the laser light is applied to the cells for the first treatment period (0.1 s), and relatively small bubbles appear near the point of action. With the accumulation of action time, up until the ninth treatment period (0.9 s), the bubbles in the cells became larger. According to the theoretical calculation, the plasma-resonant coupling of the gold nanorods at different angles to each other can enhance the photoelectric field by 1.5–60 times. GNR ensembles act as nano-lenses that are able to confine light at subwavelength dimensions, giving rise to electromagnetic field enhancements that are dozens of times of magnitude larger than those of the incident field, known as hot spots. The temperature near a hot spot rapidly rises. The organelles in the cells are expanded by the inflation gas generated by the increasing temperature to form bubbles. The intensity and distribution of the TPL changes during the laser irradiation treatment of cells, which will be discussed in detail the following sections. If the laser light is blocked from reaching the cells at this time, the bubbles will gradually disappear, but this process of generating bubbles can damage the organelles and cause the cells to become apoptotic within a certain period of time.

For comparison, a two-photon fluorescence scan at low power on cells that had not been incubated with GNRs was performed (Figure 5), and no two-photon fluorescence was found. Then, a randomly selected position in the cell was irradiated with a high-power (30 mW) laser for 1 s (Figure 5c–l). There was no change in cell morphology, indicating that the optical power acting on the cells was not sufficient to cause damage to the cells. Comparing Figure 4 with Figure 5, we found that the hot spots generated by the coupling of intracellular gold nanorod clusters are the key factors for the photothermal treatment of a single cell within a short period of time.

In order to further study the interaction between the laser and gold rods in cell culture, the cell luminescence spectra of the entire process of intracellular photothermal therapy were recorded. Figure 6c shows a single cell before laser irradiation. Figure 6d is a photograph of a cell with a laser focused onto the gold nanorod clusters. Figure 6e is an image of the cells after laser irradiation and it can be seen that the cell morphology has changed. In this process, the laser focused on the GNR clusters in the cell for 22 s, and the spectral changes in this process were recorded. Figure 6a is the two-photon fluorescence emission spectra of the cells irradiated by an 800-nm laser for 22 s. The spectral shape was similar to that of the GNRs. Figure 6b is the integration result of the spectra of Figure 6a, showing a graph of the intensity of the spectrum of a cell as a function of time during the bubble-forming process of a cell by laser excitation. It can be seen from Figure 6b that, when the laser just started irradiating the organelles containing GNR clusters in the cells, the bubble appeared. This is due to the increasing temperature inside the organelles. The field enhancement caused by the coupling of plasmon resonances between the GNRs led to this temperature increase. With the generation of the bubble, the intensity of the spectrum was greatly enhanced. As the bubble volume increased first and then decreased, the spectral intensity also decreased from strong to weak. This is a complex biophysical process. There are some possible reasons for this increase in TPL strength: first, the plasma coupling effect between the gold nanorods makes the fluorescence intensity increase; second, the organelle is thermally expanded to make the cell scattering cross-section larger; third, during the laser action process, some of the organelles change due to the heat.

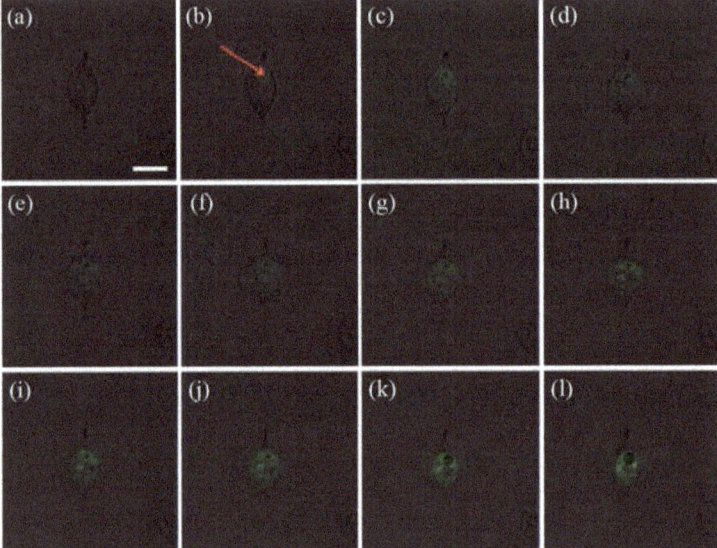

Figure 4. The image of an A375 cell incubated with GNRs for 24 h. (**a**) The bright field image before the laser treatment; (**b**) the combination of the bright field image and two-photon-absorption induced luminescence (TPL) emission image of the cell before laser treatment; (**c–l**) evolution of the cell morphology and TPL image of the excited GNR cluster when the cell was exposed to the fs laser light for different time periods of 0.1, 0.2, 0.3, 0.4, 0.5, 0.6, 0.7, 0.8, 0.9, and 1.0 s, respectively. It can be seen that afer laser treatment bubbles appears near the point of action ,with the accumulation of action time the bubbles in the cells became larger. The length of the scale bar is 20 μm.

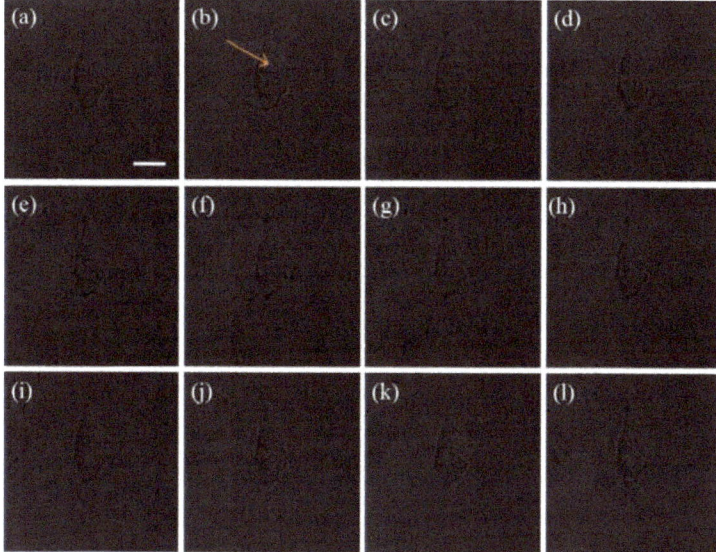

Figure 5. The image of an A375 cell not incubated with GNRs for 24 h. (**a**) The bright field image before the laser treatment; (**b**) the combination of the bright field image and TPL emission image of the cell before laser treatment; (**c**–**l**) evolution of the cell morphology and TPL image of the excited GNR cluster when the cell was exposed to the fs laser light for different time periods of 0.1, 0.2, 0.3, 0.4, 0.5, 0.6, 0.7, 0.8, 0.9, and 1.0 s, respectively. The length of the scale bar is 20 μm.

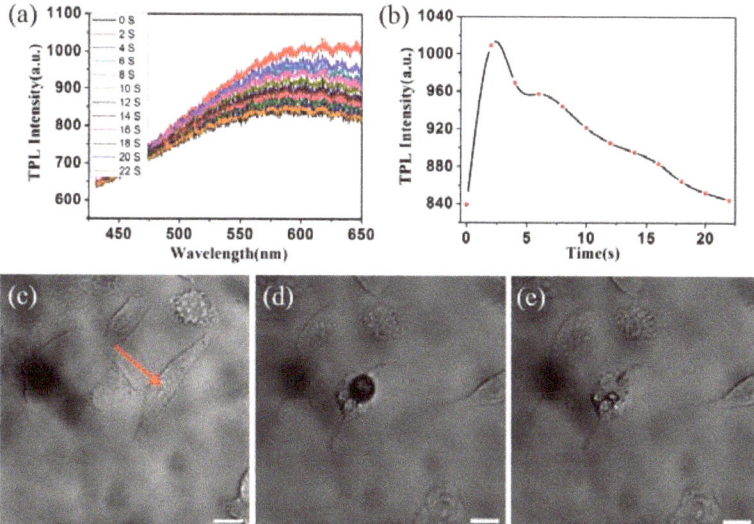

Figure 6. (**a**) Evolution of the TPL spectrum of the excited GNR cluster during laser treatment; (**b**) spectral integral intensity evolution of the TPL spectrum of the excited GNR cluster during laser treatment; A75 cells morphology before (**c**), during (**d**), and after (**e**) laser treatment. The length of the scale bar is 20 μm.

The above experimental results demonstrate that the photothermal treatment of cells can be achieved with low power using the plasma resonance coupling effect of GNR clusters. In order to further study the photothermal effect of GNRs in cells, different laser intensities were used to excite the GNRs in cells to achieve the purpose of regulating the death time of cells. Figure 7 shows the excitation of gold nanorods with different power lasers. As can be seen from Figure 7, with the 14.6-mW, laser-focused irradiation of organelles containing GNR clusters for 20 s, cells can immediately undergo necrosis. If the cells are irradiated by a 10.6-mW laser for 20 s, they do not undergo necrosis immediately, but will die after 60 min. If the cells are irradiated by a power of 9.3 mW laser for 20 s, they will die after 120 min. The results of Figure 7 are summarized in Table 1.

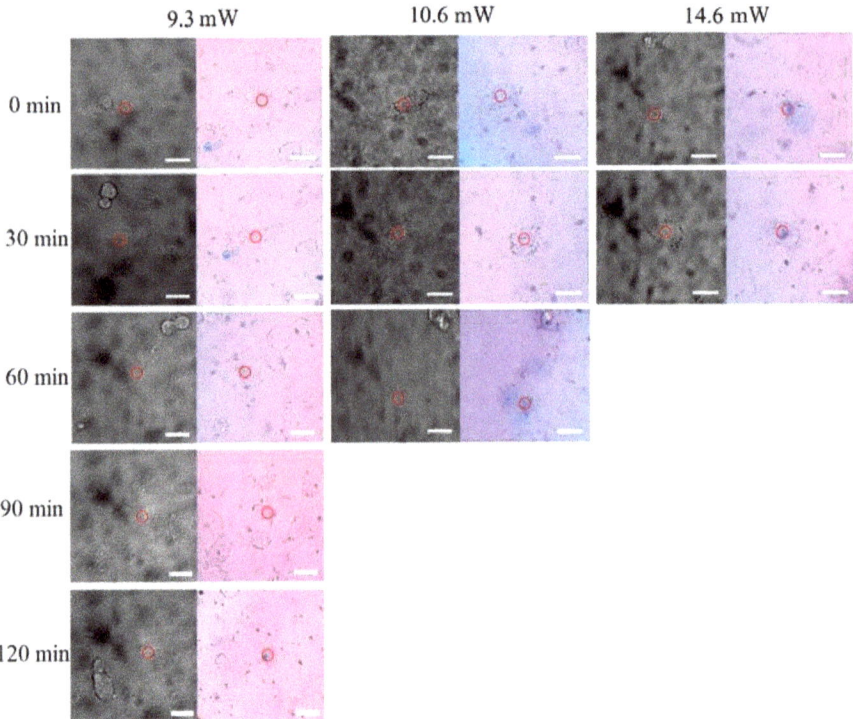

Figure 7. Bright field images of the cells on which the laser treatment was carried out. The color images show the cells dyed with Trypan Blue after the laser treatment. Laser light with different laser powers was employed in the laser treatment experiments. The cells after the irradiation of the laser light were dyed with Trypan Blue after different intervals of time of 0, 30, 60, 90, and 120 min. The blue color appearing in some areas without dead cells is caused by Trypan Blue, which did not diffuse uniformly in the experiment. The length of the scale bar is 20 µm.

Table 1. Summary of the information about the laser power and cell death time.

Power	14.6 mW	10.6 mW	9.3 mW
Cell Death Time	after 0 min	after 60 min	after 120 min

4. Conclusions

In conclusion, a fast photothermal therapy of single A375 cells was investigated in this paper. After incubation with GNRs for 24 h, the cell viability of A375 cells exceeded 80%. The cellular uptake of gold nanorods reached as high as 1360 GNRs per cell. These two conditions make gold nanorod photothermal therapy easier to develop for A375 cells. From the theoretical calculation results, it was revealed that the plasmon resonance coupling between the gold nanorods can lead to a field enhancement up to 60 times. This field enhancement leads to a larger temperature rise, which can be used to induce the death of single cells in a short time and under low laser power excitation. The position of the GNR clusters can be identified by observing the TPL images of A375 cells. Laser excitation of GNR clusters leads to changes in cell morphology and bubble generation, leading to necrosis in irradiated cells. The deaths time can be controlled by changing the input laser power. Our results help to achieve a small range of low-energy photothermal therapy, which will open up a new path for the application of nanoparticles in biomedicine.

Supplementary Materials: The following are available online at http://www.mdpi.com/2079-4991/8/11/880/s1. Figure S1: (a) The TEM image of the GNRs; (b) the statistical analysis of the GNRs, the average aspect ratio is 3.75, the standard deviation is 0.73; Figure S2: Shapes of A375 cells observed using an inverted light microscope after exposure to GNRs at a concentration of (a) 78 pM, (b) 104 pM, and (c) 0 pM (without GNRs) for 24 h; Figure S3: Experimental setup of photothermal therapy; Figure S4: The energy dispersive spectroscopy measurements of GNPs found in the TEM image of the vesicle, verifying the existence of Au element; Inset: The mass ratio of each element.

Author Contributions: Y.Y. performed the numerical simulation, conceived and designed experiments; N.Z. and X.L. performed the experiments, analyzed the data; Q.D., H.L. and Z.W. analyzed the data and wrote part of the paper; S.T. synthesized the GNRs; Y.L. designed and performed the TEM experiments, analyzed the data; H.F. and S.L. contributed the idea, structure, and design of the paper, and wrote the majority of the manuscript. All authors consulted their results and have read, critically reviewed, and agreed to the final version of the manuscript.

Funding: The authors acknowledge financial support from the National Natural Science Foundation of China (Grant No. 11674110, 61774062, and 11204092) and the Natural Science Foundation of Guangdong Province, China (Grant No. 2016A030313851).

Conflicts of Interest: The authors declare no conflict of interest.

References

1. Kang, B.; Afifi, M.M.; Austin, L.A.; El-Sayed, M.A. The nanoparticle plasmon effect: Observing drug delivery dynamics in single cells via raman/fluorescence imaging spectroscopy. *ACS Nano* **2013**, *7*, 7420–7427. [CrossRef] [PubMed]
2. Durr, N.J.; Larson, T.; Smith, D.K.; Korgel, B.A.; Sokolov, K.; Ben-Yakar, A. Two-photon luminescence imaging of cancer cells using molecularly targeted gold nanorods. *Nano Lett.* **2007**, *7*, 941–945. [CrossRef] [PubMed]
3. Zhang, W.C.; Caldarola, M.; Pradhan, B.; Orrit, M. Gold nanorod enhanced fluorescence enables single-molecule electrochemistry of methylene blue. *Angew. Chem.* **2017**, *129*, 3620–3623. [CrossRef]
4. Dagallier, A.; Boulais, E.; Boutopoulos, C.; Lachaine, R.; Meunier, M. Multiscale modeling of plasmonic enhanced energy transfer and cavitation around laser-excited nanoparticles. *Nanoscale* **2017**, *9*, 3023–3032. [CrossRef] [PubMed]
5. Lachaine, R.; Boutopoulos, C.; Lajoie, P.Y.; Boulais, É.; Meunier, M. Rational design of plasmonic nanoparticles for enhanced cavitation and cell perforation. *Nano Lett.* **2016**, *16*, 3187–3194. [CrossRef] [PubMed]
6. Chen, J.; Li, X.; Zhao, X.L.; Wu, Q.Q.; Zhu, H.H.; Mao, Z.W.; Gao, C.Y. Doxorubicin-conjugated pH-responsive gold nanorods for combined photothermal therapy and chemotherapy of cancer. *Acta Biomater.* **2018**, *3*, 347–354. [CrossRef] [PubMed]
7. Li, Y.Y.; Wen, T.; Zhao, R.F.; Liu, X.X.; Ji, T.J.; Wang, H.; Shi, X.W.; Shi, J.; Wei, J.Y.; Zhao, Y.L.; et al. Localized electric field of plasmonic nanoplatform enhanced photodynamic tumor therapy. *ACS Nano* **2014**, *8*, 11529–11542. [CrossRef] [PubMed]
8. Huschka, R.; Zuloaga, J.; Knight, M.W.; Brown, L.V.; Nordlander, P.; Halas, N.J. Light-induced release of DNA from gold nanoparticles: Nanoshells and nanorods. *J. Am. Chem. Soc.* **2011**, *133*, 12247–12255. [CrossRef] [PubMed]
9. Pan, L.; Liu, J.; Shi, J. Nuclear-targeting gold nanorods for extremely low NIR activated photothermal therapy. *ACS Appl. Mater. Interfaces* **2017**, *9*, 15952–15961. [CrossRef] [PubMed]
10. Huang, X.H.; El-Sayed, I.H.; Qian, W.; El-Sayed, M.A. Cancer cell imaging and photothermal therapy in the near-Infrared region by using Gold Nanorods. *J. Am. Chem. Soc.* **2006**, *128*, 2115–2120. [CrossRef] [PubMed]
11. Wang, H.F.; Huff, T.B.; Zweifel, D.A.; He, W.; Low, P.S.; Wei, A.; Cheng, J.X. In vitro and in vivo two-photon luminescence imaging of single gold nanorods. *PNAS* **2005**, *102*, 15752–15756. [CrossRef] [PubMed]

12. Goel, S.; Ferreira, C.A.; Chen, F.; Ellison, P.A.; Siamof, C.M.; Barnhart, T.E.; Cai, W. Activatable hybrid nanotheranostics for tetramodal imaging and synergistic photothermal/photodynamic therapy. *Adv. Mater.* **2018**, *30*, 1704367. [CrossRef] [PubMed]
13. Janani, I.; Lakra, R.; Kiran, M.S.; Korrapati, P.S. Selectivity and sensitivity of molybdenum oxide-polycaprolactone nanofiber composites on skin cancer: Preliminary in-vitro and in-vivo implications. *J. Trace Elem. Med. Biol.* **2018**, *49*, 60–71. [CrossRef] [PubMed]
14. Zhang, Z.; Wang, L.; Wang, J.; Jiang, X.; Li, X.; Hu, Z.; Ji, Y.; Wu, X.; Chen, C. Silica-coated gold nanorods as a light-mediated multifunctional theranostic platform for cancer treatment. *Adv. Mater.* **2012**, *24*, 1418–1423. [CrossRef] [PubMed]
15. Khatua, S.; Paulo, P.M.R.; Yuan, H.F.; Gupta, A.; Zijlstra, P.; Orrit, M. Resonant plasmonic enhancement of single-molecule fluorescence, by individual gold nanorods. *ACS Nano* **2014**, *8*, 4440–4449. [CrossRef] [PubMed]
16. Maltzahn, G.; Centrone, A.; Park, J.H.; Ramanathan, R.; Sailor, M.J.; Hatton, T.A.; Bhatia, S.N. SERS-coded gold nanorods as a multifunctional platform for densely multiplexed near-infrared imaging and photothermal heating. *Adv. Mater.* **2009**, *21*, 3175–3180. [CrossRef] [PubMed]
17. Soenen, S.J.; Manshian, B.; Montenegro, J.M.; Amin, F.; Meermann, B.; Thiron, T.; Cornelissen, M.; Vanhaecke, F.; Doak, S.; Parak, W.J.; et al. Cytotoxic effects of gold nanoparticles: A multiparametric Study. *ACS Nano* **2012**, *6*, 5767–5783. [CrossRef] [PubMed]
18. Zhang, X.D.; Wu, D.; Shen, X.; Chen, J.; Sun, Y.M.; Liu, P.X.; Liang, X.J. Size-dependent radiosensitization of PEG-coated gold nanoparticles for cancer radiation therapy. *Biomaterials* **2012**, *33*, 6408–6419. [CrossRef] [PubMed]
19. Alkilany, A.M.; Nagaria, P.K.; Hexel, C.R.; Shaw, T.J.; Murphy, C.J.; Wyatt, M.D. Cellular uptake and cytotoxicity of gold nanorods: Molecular origin of cytotoxicity and surface effects. *Small* **2009**, *5*, 701–708. [CrossRef] [PubMed]
20. Wang, C.; Chen, J.; Talavage, T.; Irudayaraj, J. Gold Nanorod/Fe_3O_4 Nanoparticle "Nano-Pearl-Necklaces" for Simultaneous Targeting, Dual-Mode Imaging, and Photothermal Ablation of Cancer Cells. *Angew. Chem. Int. Ed.* **2009**, *48*, 2759–2763. [CrossRef] [PubMed]
21. Marasini, R.; Pitchaimani, A.; Nguyen, T.D.; Comer, J.; Aryal, S. Influence of Polyethylene Glycol Passivation on the Surface Plasmon Resonance Induced Photothermal Properties of Gold Nanorods. *Nanoscale* **2018**, *10*, 13684–13693. [CrossRef] [PubMed]
22. Zhang, Y.; Zhan, X.; Xiong, J.; Peng, S.; Huang, W.; Joshi, R.; Cai, Y.; Liu, Y.; Li, R.; Yuan, K.; et al. Temperature-dependent cell death patterns induced by functionalized gold nanoparticle photothermal therapy in melanoma cells. *Sci Rep.* **2018**, *8*, 8720. [CrossRef] [PubMed]
23. Ali, M.R.K.; Snyder, B.; El-Sayed, M.A. Synthesis and optical roperties of small Au nanorods using a seedless growth technique. *Langmuir* **2012**, *28*, 9807–9815. [CrossRef] [PubMed]
24. Qiu, Y.; Liu, Y.; Wang, L.M.; Xu, L.G.; Bai, R.; Ji, Y.L.; Wu, X.C.; Zhao, Y.L.; Li, Y.F.; Chen, C.Y. Surface chemistry and aspect ratio mediated cellular uptake of Au nanorods. *Biomaterials* **2010**, *31*, 7606–7619. [CrossRef] [PubMed]
25. Fan, H.H.; Le, Q.; Lan, S.; Liang, J.X.; Tie, S.L.; Xu, J.L. Modifying the mechanical properties of gold nanorods by copperdoping and triggering their cytotoxicity with ultrasonic wave. *Colloids Surf. B Biointerfaces* **2018**, *163*, 47–54. [CrossRef] [PubMed]
26. Zhou, W.B.; Liu, X.S.; Ji, J. More efficient NIR photothermal therapeutic effect from intracellular heating modality than extracellular heating modality: An in vitro study. *J. Nanopart. Res.* **2012**, *14*, 1128–1144. [CrossRef]
27. Yee, K.S. Numerical solution of inital boundary value problems involving Maxwell's equations in isotropic media. *IEEE. Trans. Antenn. Propag.* **1966**, *14*, 302–307.
28. Yang, P.; Liou, K.N. Finite-difference time domain method for light scattering by small ice crystals in threedimensional space. *J. Opt. Soc. Am. A* **1996**, *13*, 2072–2085. [CrossRef]
29. Rubio, G.G.; Izquierdo, J.G.; Bañares, L.; Tardajos, G.; Rivera, A.; Altantzis, T.; Bals, S.; Rodríguez, O.P.; Martínez, A.G.; Marzaín, L.M.L. Femtosecond laser-controlled tip-to-tip assembly and welding of gold nanorods. *Nano Lett.* **2015**, *15*, 8282–8288. [CrossRef] [PubMed]

30. Lee, A.; Andrade, G.F.S.; Ahmed, A.; Souza, M.L.; Coombs, N.; Tumarkin, E.; Liu, K.; Gordon, R.; Brolo, A.G.; Kumacheva, E. Probing dynamic generation of hot-spots in self-assembled chains of gold nanorods by surface-enhanced raman scattering. *J. Am. Chem. Soc.* **2011**, *133*, 7563–7570. [CrossRef] [PubMed]
31. Lee, A.; Ahmed, A.; dos Santos, D.P.; Coombs, N.; Park, J., II; Gordon, R.; Brolo, A.G.; Kumacheva, E. Side-by-side assembly of gold nanorods reduces ensemble-averaged SERS intensity. *J. Phys. Chem. C* **2012**, *116*, 5538–5545. [CrossRef]

© 2018 by the authors. Licensee MDPI, Basel, Switzerland. This article is an open access article distributed under the terms and conditions of the Creative Commons Attribution (CC BY) license (http://creativecommons.org/licenses/by/4.0/).

Article

Biosynthesis of Silver Nanoparticles Using *Ligustrum Ovalifolium* Fruits and Their Cytotoxic Effects

Bianca Moldovan [1], Vladislav Sincari [2], Maria Perde-Schrepler [3] and Luminita David [1,*]

[1] Research Center for Advanced Chemical Analysis, Instrumentation and Chemometrics (ANALYTICA), Faculty of Chemistry and Chemical Engineering, Babeş-Bolyai University, 11 Arany Janos Street, Cluj-Napoca 400028, Romania; bianca@chem.ubbcluj.ro
[2] Faculty of Chemistry and Chemical Engineering, Babeş-Bolyai University, 11 Arany Janos Street, Cluj-Napoca 400028, Romania; shinkar.vlad@mail.ru
[3] "Ion Chiricuta" Oncology Institute, 34–36 Republicii Street, Cluj-Napoca 400015, Romania; pmariaida@yahoo.com
* Correspondence: muntean@chem.ubbcluj.ro; Tel.: +40-264-593-833

Received: 16 July 2018; Accepted: 17 August 2018; Published: 18 August 2018

Abstract: The present study reports for the first time the efficacy of bioactive compounds from *Ligustrum ovalifolium* L. fruit extract as reducing and capping agents of silver nanoparticles (AgNPs), developing a green, zero energetic, cost effective and simple synthesis method of AgNPs. The obtained nanoparticles were characterized by UV-Vis spectroscopy, transmission electron microscopy (TEM), X-ray diffraction (XRD) and Fourier Transform Infrared spectroscopy (FTIR), confirming that nanoparticles were crystalline in nature, spherical in shape, with an average size of 7 nm. The FTIR spectroscopy analysis demonstrated that the AgNPs were capped and stabilized by bioactive molecules from the fruit extract. The cytotoxicity of the biosynthesized AgNPs was in vitro evaluated against ovarian carcinoma cells and there were found to be effective at low concentration levels.

Keywords: phytosynthesis; silver nanoparticles; *Ligustrum ovalifolium* L.; cytotoxic activity; ovarian carcinoma cells

1. Introduction

Nanostructures of noble metals were lately immensely investigated due to their remarkable physical and chemical properties. The beneficial effects of silver salts have been noticed since antiquity. Reducing the particle size of materials is an efficient and reliable tool to improve their biocompatibility. Nanoparticles can be synthesized by several ways, such as physical, chemical or biological methods. Silver nanoparticles can be obtained by various chemical and photochemical reduction reactions, by thermal decomposition, by electrochemical methods, radiation or sonochemical assisted synthesis [1]. All these processes are efficient techniques to synthesize silver nanoparticles but they also have some drawbacks. The physical and chemical processes are expensive and use hazardous chemicals which may generate important environmental problems and can require a great deal of energy [2]. The as synthesized silver nanoparticles are chemically contaminated and require an advanced purification especially when they are intended to be used for medical applications. The biological methods are environmental friendly, cost effective and easily scaled up for large scale synthesis of nanoparticles and involve microorganisms, enzymes or plant extracts [3–5]. Various recent studies demonstrated the efficacy of fruit extracts such as *Acacia nilotica*, *Phoenix dactylifera*, *Tamarindus indica*, *Sambucus nigra*, *Licium barbarum* in the synthesis of silver nanoparticles [6–9]. The phytochemical compounds present in fruits such as flavonoids, carotenoids, aldehydes, ketones, proteins and carboxylic acids may act as bioreducing agents for Ag ions to silver nanoparticles. Metal nanoparticles obtained by phytomediated green synthetic methods combine the biological effects of metal and bioactive molecules present in the plant extract which are responsible for the reduction and stabilization of the nanoparticles, so they can be used as reliable tools in the field of nanomedicine [7,10–12]. Plant mediated synthesized nanoparticles have also the advantage of being safer for biomedical purposes as microbe or chemical mediated synthesized nanoparticles [13,14].

In the recent years, several biomedical applications have been reported for silver nanoparticles [15–17]. Since the ancient times, silver has been used in wound healing and in the 19th century its antimicrobial activity was established, this being the most well-known and exploited biological application of silver nanoparticles. Apart their antibacterial activity, AgNPs have been also proved as efficient antifungal and antivirucidal agents (inhibit HIV, Takaribe virus, hepatitis B, A/H1N1 virus) [16,18]. Recent publications reported the potential therapeutic applications of silver nanoparticles in cancer and inflammatory diseases [4,6,10,16].

Ligustrum ovalifolium L. is commonly called California privet or garden privet, is an ornamental semi-evergreen shrub original from East Asia, widely cultivated as ornamental plant. *Ligustrum* (privet) fruits are known to contain phenolic acids, flavonoids and triterpenoids, responsible for their antihyperglycemic, anticarcinogenic effect and immunomodulatory activity [3,19–21].

Traditional Chinese medicine uses privet fruits as tonic for liver and kidneys [17]. Modern medicine recorded the extract of these fruits to possess immunomodulatory, anti-inflammatory, antitumor and anti-ageing effects [22]. *Ligustrum lucidum* fruits exhibit antiproliferative activity against lung, breast, liver, pancreatic and colorectal carcinoma cells [23–25].

Ovarian carcinoma is one of the leading primary causes of cancer-related fatality in women [26]. Therefore, finding new therapeutic agents to fight against the proliferation of these carcinoma cells is of great concern.

The objective of the present work was to develop a phytomediated green synthesis method of silver nanoparticles, without using any environmental deleterious chemical reducing or capping agents such as sodium borohydride, Tollens reagent, N,N-dimethyl formamide and polyvinyl alcohol (PVA) [27], by exploiting the antioxidant effects of compounds present in the *Ligustrum ovalifolium* L. fruit extract and to investigate their cytotoxicity against A2780 ovarian carcinoma cells.

2. Materials and Methods

2.1. Reagents

Cell titre blue reagent was purchased from Promega (Darmstadt, Germany). Cell lines and all other chemicals and reagents were purchased from Sigma-Aldrich (Darmstadt, Germany) and were of analytical purity.

2.2. Preparation of the Extract

Garden privet fruits were harvested in September 2017 from Cluj-Napoca, Romania. To 2.5 g of fresh milled fruits, 50 mL of distilled water were added and the mixture was stirred for 1 h at room temperature and then filtered.

2.3. Determination of Total Phenolic Content

The total phenolic content of the garden privet fruits extract was colorimetrically determined using the Folin-Ciocalteu method [28]. Aliquots of 250 µL extract were transferred into a test tube (each analysis was conducted in triplicate) and then mixed with 1.5 mL of Folin-Ciocalteu reagent. The samples were incubated in the dark for 5 minutes and then 1.2 mL of sodium carbonate solution (0.7 M) was added and thoroughly mixed. After standing 2 hours at room temperature in the dark, the absorbance of the resulting blue solutions was measured at 765 nm, using a double beam UV-Vis spectrophotometer (Perkin-Elmer Lambda 25, PerkinElmer Inc., Waltham, MA, USA). The total phenolic content was calculated and expressed as gallic acid equivalents (mg GAE/L) using a calibration curve of the gallic acid standard solution (0–100 µg/mL).

2.4. Determination of Antioxidant Activity

The free radical scavenging activity was determined using the ABTS (2,2′-azinobis-3-ethyl-benzthiazino-6-sulphonic acid) method [29]. The ABTS$^{+\cdot}$ stock solution was obtained by mixing equal volumes of 7mM ABTS solution with an oxidant agent (2.45 mM potassium persulfate solution). The stock solution was diluted to give an absorbance between 0.6–0.8 at 734 nm. Aliquots of 50 µL fruit extract were added to 3 mL diluted ABTS$^{+\cdot}$ solution and the absorbencies of the samples were red at 734 nm exactly after 15 minutes. The antioxidant activity was calculated using a calibration curve of Trolox standard (0–400 µmol/L Trolox) and expressed as µM Trolox equivalents.

2.5. Synthesis of AgNPs

The green synthesis of silver nanoparticles was achieved by adding 5 mL of *Ligustrum ovalifolium* L. fruit extract to 25 mL 1 mM aqueous silver nitrate solution under stirring at room temperature for the bioreduction of Ag$^+$ ions to Ag^0. The effect of the pH of the reaction medium on the silver nanoparticles formation and characteristics was investigated over a pH range from 6 to 10. The influence of the pH value on the formation of AgNPs was spectrophotometrically monitored. The pH of the resulting solution was adjusted by adding 1% aqueous NaOH solution. The rapid formation of a yellowish-brown colour indicated the reduction of the silver ions. After 3 hours under constant stirring, the resulting solutions were centrifuged at 10,000 rpm for 15 minutes. The separated nanoparticles were washed with double distilled water and dried. Three mg solid AgNPs were obtained.

The influence of the ratio of silver ions to the fruits extract which act as reducing agent was investigated at optimized pH value. Different amounts of plant extract were added to 25 mL 1mM silver nitrate solution in order to achieve the desired ratios (1:1; 1:3; 1:7; 1:10; 1:12 and 1:15 respectively).

2.6. Characterization of the AgNPs

The formation and the stability of silver nanoparticles were spectrophotometrically monitored, using a Perkin Elmer Lambda 25 UV-Vis spectrophotometer. The morphological characteristics of the synthesized nanoparticles were carried out using transmission electron microscopy (H-7650 120 kV Automatic transmission electron microscopy (TEM), Hitachi, Tokyo, Japan). In order to identify the biomolecules responsible for capping and stabilizing the silver nanoparticles, these were analysed by Fourier transform infrared spectroscopy (FTIR) using a Bruker Vector 22 FT-IR spectrometer, Bruker, Dresden, Germany). The obtained AgNPs were subjected to X-Ray diffraction analysis, data being recorded on a D8 Advance X-ray diffractometer with CuK$_\alpha$1 radiation.

The thermogravimetric analysis (TGA) of the biosynthesized AgNPs obtained using *Ligustrum* fruit extract was realised in nitrogen atmosphere using a thermogravimetric analyser (SDT Q600, TA Instruments, New Castle, DE, USA) in a temperature range 40–1000 °C, at a heating rate of 10°/min.

A Malvern Zetasizer Nanoseries compact scattering spectrometer (Malvern instruments Ltd., Worcestershire, UK) was used to determine the zeta potential of the synthesized silver nanoparticles. The hydrodynamic diameter of particles was determined in aqueous solution also by the Zetasizer.

2.7. Cytotoxic Activity

The synthesized AgNPs were tested on two cell lines: A2780 and A2780-Cis (cisplatin resistant), two human ovarian carcinoma cell lines. The cells were maintained in specific media: Roswell Park Memorial Institute (RPMI), supplemented with 10% foetal calf serum, glutamine and antibiotics. The cells were kept at 37 °C in a humid incubator with 5% CO$_2$.

The AgNPs cytotoxicity was evaluate by a fluorometric method, using cell titre blue. This is a homogeneous fluorometric method which estimates the number of viable cells which maintain their metabolic capacity to reduce resazurin (dark blue colour) to resofurin (pink-purple). Resofurin is highly fluorescent and can be quantified with a fluorescence plate reader at 579 nm extinction and 584 nm emission.

Cells were plated in 96-well plates at a density of 20,000 cells/well. The cells were treated with different concentrations of AgNPs solutions: 180; 90; 45; 22.5; 9; 4.5; 2.25; 0.9; 0.01 µg/mL. The cells were also treated with serial dilutions of the *Ligustrum ovalifolium* L. extract. From the initial concentration of the extract: 190 µg GAE/mL, the following concentrations were obtained: 170; 142.5; 95; 47.5; 19; 9.5; 4.75; 1.9 µg/mL. As a positive control, the cytotoxicity of cisplatin on the ovarian carcinoma cell lines was tested. After 24h incubation, 20 µL cell titre blue reagent was added to 100 µL culture medium in each well. After 1h incubation the fluorescence was recorded. At each concentration, the surviving fraction was calculated as fluorescence of the sample/fluorescence of control (non-treated cells) × 100. Each experiment was done in triplicate.

Statistical analysis was performed using GraphPad Prism 5 software (San Diego, CA, USA). Inhibitory concentration (IC50) values were calculated using the non-linear regression. Comparisons were made using one-way ANOVA, considering $p < 0.05$ as criteria of significance.

3. Results and Discussion

3.1. Synthesis and Characterization of Silver Nanoparticles

Oleuropein, luteolin and ligstroside are the main flavonoids present in privet fruits [30], compounds which confer these fruits a high antioxidant capacity and make them suitable candidates for the reduction of metallic ions to metallic nanoparticles. The antioxidant capacity of the privet fruit extract was determined using the ABTS assay and was found to be 340 µM Trolox. The total phenolic content of the extract, assessed by the Folin Ciocalteu method, was 190 µg GAE/mL.

The bioactive molecules present in the *Ligustrum ovalifolium* L. fruit extract served as reducing and capping agents for the synthesis and stabilization of silver nanoparticles. The colour change of the synthesis mixture to a yellowish brown colour, due to the surface plasmon resonance (SPR) of the silver nanoparticles in the solution, indicated the bioreduction of the silver ions [31]. The UV-Vis spectra of the silver nanoparticles obtained at different pH values are presented in Figure 1. The SPR peaks of AgNPs appeared between 402 and 413 nm, in accordance to the characteristic SPR band of silver nanoparticles, typically in the range of 400–500 nm [32]. The absorption peak appeared at 413 nm in acidic conditions (pH = 6). Increasing the pH value of the reaction media, the SPR band of silver nanoparticles was shifted to lower wavelengths (402 nm at pH = 7) suggesting the formation of smaller in diameter AgNPs [33].

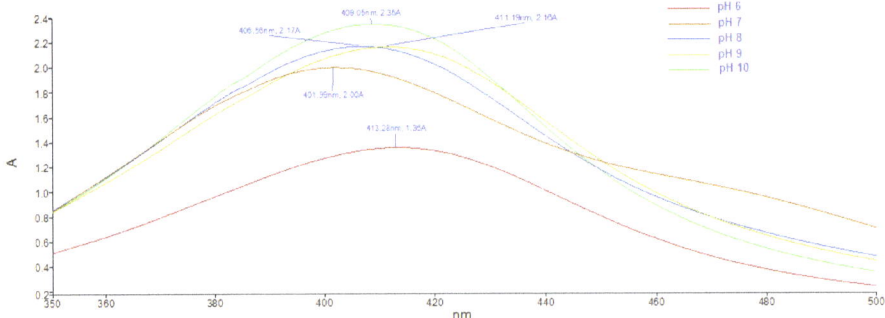

Figure 1. UV-Vis spectra of silver nanoparticles at different pH values.

The size and shape of the phytosynthesized silver nanoparticles were monitored by TEM analysis. In acidic conditions, silver nanoparticles trend to aggregate whereas alkaline pH of the reaction medium led to formation of a large number of silver nanoparticles with smaller diameter. The TEM images (Figure 2) indicate the formation AgNPs, due to the stabilization by the biomolecules present in the *Ligustrum ovalifolium* L. fruit extract. In acidic conditions (pH = 6) the synthesized silver nanoparticles show irregular but mostly spherical shape morphology (Figure 2a) with sizes in the range of 25 nm. In order to realize the histograms for size determination of silver nanoparticles over 100 particles in random field of TEM images were measured. The particles obtained at pH = 7 are smaller in size (in the range of 10 nm Figure 2b) with more regular shape but they seem agglomerated. In alkaline conditions (pH = 8 to 10) the influence of the pH value on the size and morphology of the phytosynthesized silver nanoparticles was not significant (Figure 2c–e). The size of the spherical AgNPs obtained in these conditions was in the range of 6–13 nm with an average diameter of 7 nm (at pH = 10). Taking all these facts into account, the optimum pH value for the synthesis of *Ligustrum ovalifolium* fruit extract mediated silver nanoparticles is 10, value at which highly dispersed, spherical, regular in shape and small in diameter AgNPs were obtained.

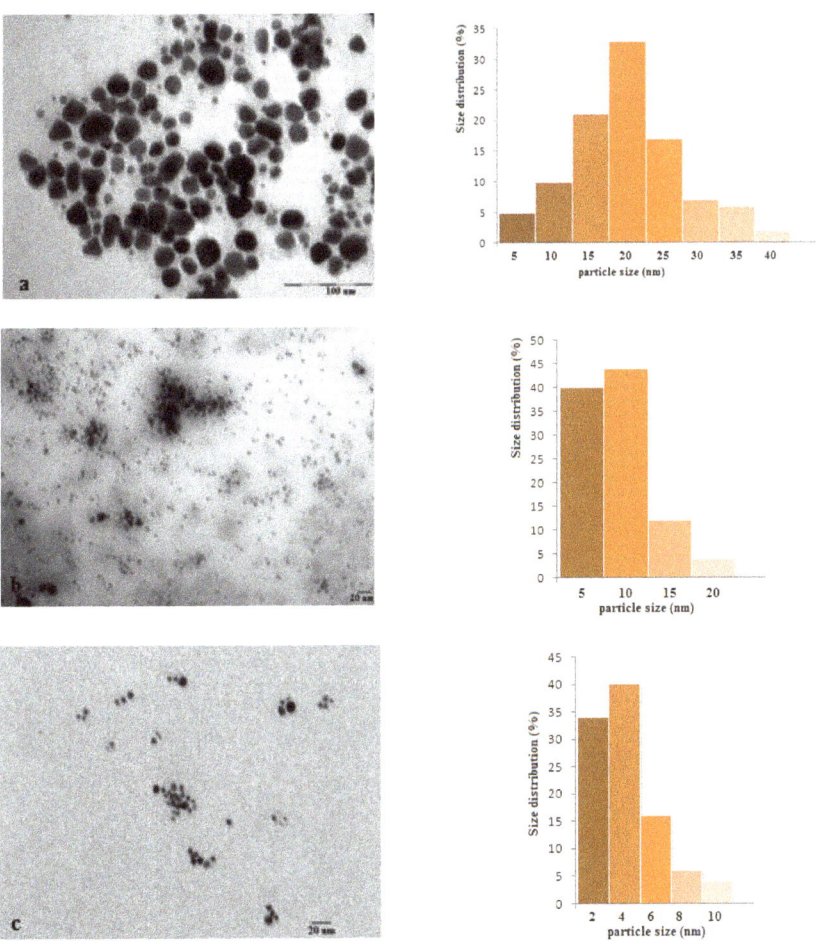

Figure 2. *Cont.*

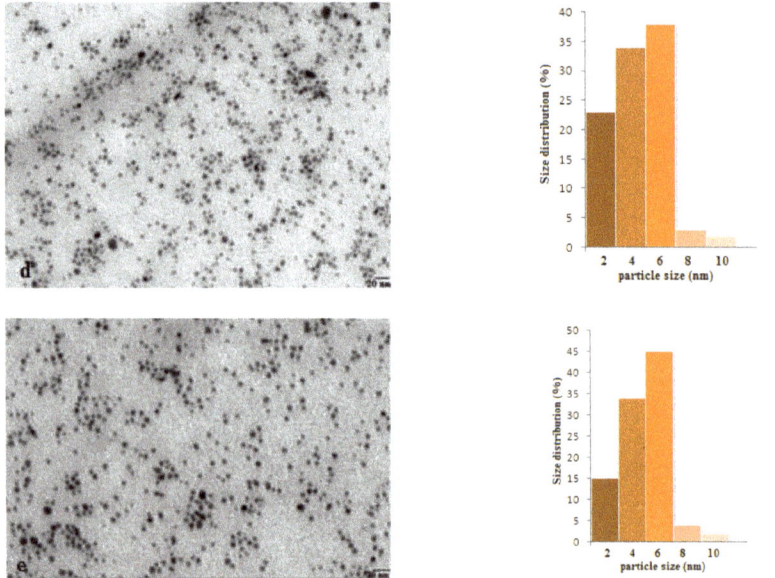

Figure 2. Transmission electron microscopy (TEM) images and corresponding histograms of silver nanoparticles at different pH values: (**a**) pH = 6; (**b**) pH = 7; (**c**) pH = 8; (**d**) pH = 9 and (**e**) pH = 10.

The size, the shape and the yield of nanoparticles synthesis is also affected by the ratio of reducing agent to the metal ions. Various ratios: 1:1; 1:3; 1:7; 1:10; 1:12 and 1:15 respectively were investigated at the previously optimized pH value (pH = 10), in order to enhance the yield of AgNPs formation. The maximum absorbance in the UV-Vis spectra (Figure 3) was obtained for a 1:3 ratio fruit extract to silver nitrate solution.

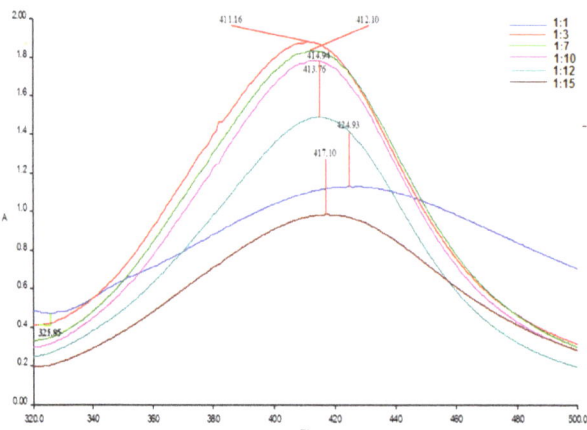

Figure 3. UV-Vis spectra of silver nanoparticles at different ratios fruit extract: silver nitrate solution.

Moreover, the AgNPs obtained at this ratio exhibited surface plasmon resonance on a narrow absorption range, fact that indicates the monodispersity of the silver nanoparticles. By increasing the ratio to 1:15, the absorption maximum was shifted to a higher λ_{max} value, indicating a larger size of the formed nanoparticles and also a reduction in the absorption was observed, which may be due to aggregation of the AgNPs. The optimum ratio required for the synthesis of AgNPs with privet fruit extract was found to be 1:3.

The crystalline nature of the AgNPs was confirmed by the X-ray diffraction pattern (Figure 4). The X-ray diffractogram presents four prominent peaks at 2θ values of 37.59°, 44.07°, 64.17° and 77.22° corresponding to the (111), (200), (220) and (311) Bragg's reflections assigned to lattice planes of face centred cubic (FCC) structure of metallic silver, in good agreement to reference of FCC structure of Joint Committee of Powder Diffraction Standard (JCPDS) file No. 04-0783.

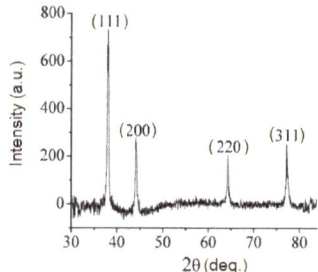

Figure 4. X-Ray Diffraction (XRD) pattern of silver nanoparticles.

Fourier transform infrared spectroscopy (FTIR) analysis was accomplished in order to identify the biomolecules from the fruit extract responsible for the reduction of silver ions and for the efficient stabilization of the silver nanoparticles. Figure 5 presents the FTIR spectra of the *Ligustrum ovalifolium* fruit extract and of the green synthesized AgNPs. The FTIR spectra of the fruit extract and silver nanoparticles present similar absorption bands. The FTIR analysis confirmed the dual role of the fruit extract as reducing and capping agent. The broad band at 3296 cm^{-1} is due to the O-H stretching vibration attributed to the polyphenols from the fruit extract. The FTIR spectrum of the silver nanoparticles shows several shifts of the absorption bands and also some intensity changes of these bands, compared to the spectrum of the fruit extract. The bands appearing at 2933, 1560 and 1383 and 1078 cm^{-1} indicate the presence of different functional groups and were assigned to C-H stretching vibration, C=C stretching vibration from aromatic rings and C-O stretching or O-H deformation vibration, respectively [34]. Most of the IR bands probably arise from the flavonoids anthocyanins and phenolic compounds from the fruit extract, indicating that the phytosynthesized AgNPs were stabilized by these biomolecules [35,36].

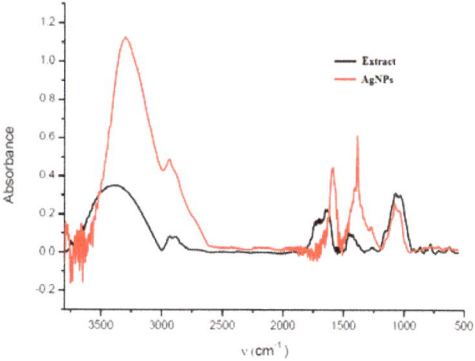

Figure 5. Fourier transform infrared (FTIR) spectra of fruit extract and silver nanoparticles.

The observed peaks in the FTIR spectrum clearly indicate that the bioactive compounds from *Ligustrum ovalifolium* L. fruit extract are present as coating/capping agents at the surface of the silver nanoparticles.

The presence of the capping organic biomolecules on the surface of the privet extract synthesized silver nanoparticles was also confirmed by the thermogravimetric analysis (TGA). Figure 6 presents the TGA pattern of the green synthesized silver nanoparticles. The TGA curve clearly indicates an initial weight loss observed at around 100 °C, due to the loss of bound and unbound water molecules present in the obtained silver nanoparticles. A second weight loss appeared while temperature increases at 150–340 °C which accounted a 27% of the total AgNPs weight. This weight loss can be attributed to the degradation of organic compounds such as phenolic acids, flavonoids and carbohydrates [37]. A steady loss of weight appeared until 1000 °C which accounted for 29%, probably determined by the thermal degradation of resistant aromatic compounds. The observed behaviour in the TGA is a consequence of the thermal degradation of organic compounds from the fruit extract present in the nanoparticles powder which can be estimated to be 56% from the phytosynthesized AgNPs.

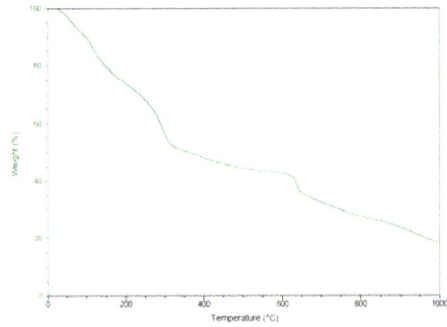

Figure 6. Thermogravimetric analysis (TGA) of silver nanoparticles.

The zeta potential of the green synthesized AgNPs was also determined using dynamic light scattering (DLS). The value of zeta potential gives important information about the surface of the silver nanoparticles and their stability. Generally, values of zeta potential of a colloidal suspension indicate the stability of the nanoparticles from solution [38]. Figure 7 presents the zeta potential (Figure 7a) and the DLS size distribution (Figure 7b) of the aqueous suspension of privet fruit extract mediated synthesized silver nanoparticles.

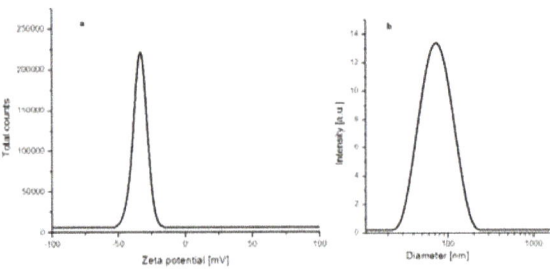

Figure 7. (a) Zeta potential. (b) hydrodynamic diameter of the obtained AgNPs.

The value of the determined zeta potential was −31.8 mV, suggesting a high stability of the silver suspension. The negative value of the zeta potential once more confirms that the organic phytocomponents from the fruit extract, eventually with a phenolic structure, acted as capping agent and stabilized the AgNPs. The determined average particle hydrodynamic size was 91.8 nm, similar to that reported by other studies in which also phytosynthesized silver nanoparticles were obtained [39]. As expected, the particles size, in aqueous solution, obtained by DLS was larger than that obtained from TEM, due to the different principles used for analysis.

3.2. Cytotoxic Activity

The silver nanoparticles are known to inhibit cancer cells growth and thus could be potential anti-cancer therapeutic agents. Previous studies report the ability of AgNPs to inhibit growth of various cancer cell lines such as breast, ovarian, lung, glioblastoma, hepatic [40].

In vitro cytotoxicity of the biologically synthesized silver nanoparticles was investigated against two human ovarian cancer cell lines, A2780 and A2780Cis (cisplatin resistant) at different concentrations. The cells viability gradually decreased by increasing the concentration of AgNPs from 0.01 to 180 µg/mL (Figure 8) indicating a dose-dependent cytotoxic effect of the nanoparticles. IC50 values at 24 h incubation were calculated using nonlinear regression and four-parameter sigmoidal curve fit, with each point representing mean ±SEM in three separate measurements. The concentrations of the nanoparticles that reduced cell viability by 50% (IC50) were: 7 µg/mL for A2780 and 14.04 µg/mL for A2780-Cis, respectively.

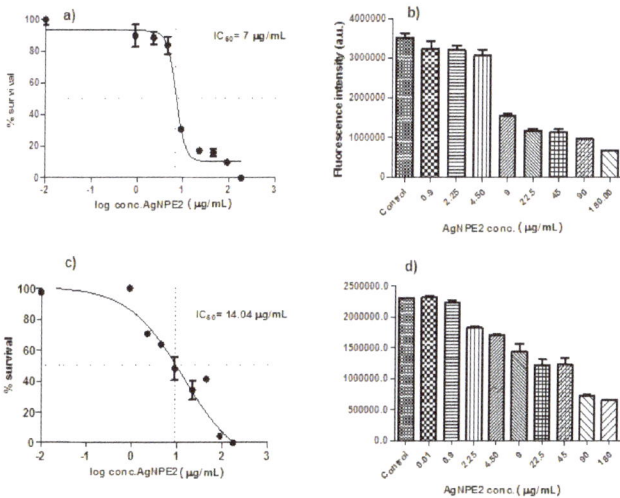

Figure 8. Dose-dependent toxicity of AgNPs on ovary carcinoma cell lines: (**a**) A2780; (**c**) A2780-Cis. The concentration of AgNPs for the significant reduction of cells viability: (**b**) A2780 (9 µg/mL); (**d**) A2780-Cis (2.25 µg/mL).

Our results are in agreement with those obtained in previous studies. Fahrenholtz et al. [41] determined IC50 value of cca. 7.2 µg/mL against A2780 cells when exposed to polyvinylpyrrolidone (PVP) coated silver nanoparticles but the advantage of using green synthesized nanoparticles could be the absence of a potential toxic synthetic capping agent. Young and co-workers [42] evaluated the influence of surface coating on the toxicity of silver nanoparticles in model organism *Caenorhabditis elegans*. They investigated citrate-coated, PVP-coated and gum arabic-coated AgNPs and demonstrated that the capping agent affects the degree of toxicity of silver nanoparticles. Gum arabic-coated (EC50 = 0.9 µM) have been proved to be 9-fold more toxic than PVP-coated (EC50 = 8 µM) which in turn were 3-fold more toxic than citrate-coated AgNPs (EC50 = 31 µM), all these differences being attributed to differences in toxicity of the surface coating agent.

Lakshmanan and co-workers [43] also tested the cytotoxicity of phytosynthesized AgNPs on ovarian cancer cell lines but the obtained IC50 value was 30 μg/mL which is cca. 4-fold higher than the IC50 concentration of *Ligustrum ovalifolium* fruit extract mediated synthesized silver nanoparticles obtained in this study.

The extract showed low toxicity against the ovarian carcinoma cell lines, their viability being reduced significantly starting with the 95 μg/mL concentration for A2780 cells ($p < 0.001$) and 9.5 μg/mL ($p < 0.5$) and 47.5 ($p < 0.001$) for the cisplatin resistant cells (one-way ANOVA, Dunnett's multiple comparisons test). The viability of both cell lines was reduced with 50% only when they were treated with the undiluted extract (no growth media) (Figure 9a,b). These results show that the extract has no/reduced toxicity towards the ovarian carcinoma cell lines.

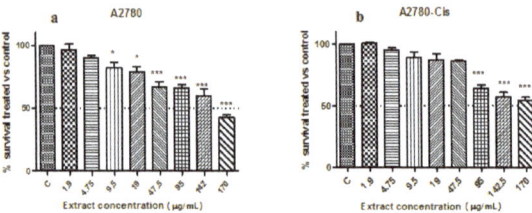

Figure 9. The viability of ovarian carcinoma cell lines treated with *Ligustrum ovalifolium* fruit extract: (**a**) A2780 and (**b**) A2780-Cis.

The inhibitory effect on cancer cell growth enabled us to anticipate the promising anticancer potential of the obtained silver nanoparticles against ovarian cancer.

4. Conclusions

The present study presents for the first time an eco-friendly, cost effective, efficient and simple method for the biosynthesis of AgNPs using bioactive compounds from the aqueous extract of *Ligustrum ovalifolium* L. fruits as reducing and capping agents. The synthesized nanoparticles proved potential cytotoxic effect against ovarian carcinoma cells and may be suitable for biomedical applications as promising alternative therapeutic agents for cancer.

Author Contributions: L.D. designed the experiment; B.M., V.A. and M.P.-S. performed the experimental work; B.M. and L.D. contributed to the writing of the paper.

Funding: This research was funded by the Ministry of National Education and Scientific Research, Romania as a part of the research project No. PN-III-P4-ID-PCE-2016-0396

Conflicts of Interest: The authors declare no conflict of interest. The funders had no role in the design of the study; in the collection, analyses, or interpretation of data; in the writing of the manuscript and in the decision to publish the results.

References

1. Janardhanan, R.; Karuppaiah, M.; Hebalkar, N.; Rao, T.N. Synthesis and surface chemistry of nano silver particles. *Polyhedron* **2009**, *28*, 2522–2530. [CrossRef]
2. Thakkar, N.; Mhatre, S.S.; Parich, R.Y. Biological synthesis of metallic nanoparticles. *Nanomed. Nanotechnol. Biol. Med.* **2010**, *6*, 257–262. [CrossRef] [PubMed]
3. Wang, Z.H.; Hsu, C.C.; Yin, M.C. Antioxidative characteristics of aqueous and ethanol extracts of glossy privet fruit. *Food Chem.* **2009**, *112*, 914–918. [CrossRef]
4. Moldovan, B.; David, L.; Vulcu, A.; Olenic, L.; Perde-Schrepler, M.; Fischer-Fodor, E.; Baldea, I.; Clichici, S.; Filip, G.A. In vitro and in vivo anti-inflammatory properties of green synthesized silver nanoparticles using *Viburnum opulus* L. fruit extract. *Mater. Sci. Eng. C* **2017**, *79*, 720–727. [CrossRef] [PubMed]

5. Opris, R.; Tatomir, C.; Olteanu, D.; Moldovan, R.; Moldovan, B.; David, L.; Nagy, L.; Decea, N.; Kiss, M.; Filip, G.A. The effect of *Sambucus nigra* L. extract and phytosynthesized gold nanoparticles on diabetic rats. *Colloids Surf. B* **2017**, *150*, 192–200. [CrossRef] [PubMed]
6. Mohammed, A.E.; Al-Qahtani, A.; al-Mutairi, A.; Al-Shamri, B.; Aabed, K. Antibacterial and cytotoxic potential of biosynthesized silver nanoparticles by some plant extracts. *Nanomaterials* **2018**, *8*, 382. [CrossRef] [PubMed]
7. Moldovan, B.; David, L.; Achim, M.; Clichici, S.; Filip, G.A. A green approach to phytomediated synthesis of silver nanoparticles using *Sambucus nigra* L. fruit extract and their antioxidant activity. *J. Mol. Liq.* **2016**, *221*, 271–278. [CrossRef]
8. Jayaprakash, N.; Vijaya, J.J.; Kaviyarasu, K.; Kombaiah, K.; Kennedy, L.J.; Ramalingam, R.J.; Munusamy, M.A.; Al-Lohedan, H.A. Green synthesis of Ag nanoparticles using Tamarind fruit extract for the antibacterial studies. *J. Photochem. Photobiol. B Biol.* **2017**, *169*, 178–185. [CrossRef] [PubMed]
9. Dong, C.; Cao, C.; Zhang, X.; Zhan, Y.; Wang, X.; Yang, X.; Zhou, K.; Xiao, X.; Yuan, B. Woolfberry fruit (*Licium barbarum*) extract mediated novel route for the green synthesis of silver nanoparticles. *Opt. Int. J. Light Electron Opt.* **2017**, *130*, 162–170. [CrossRef]
10. Filip, A.G.; Potara, M.; Florea, A.; Baldea, I.; Olteanu, D.; Bolfa, P.; Clichici, S.; David, L.; Moldovan, B.; Olenic, L.; et al. Comparative evaluation by scanning confocal Raman spectroscopy and transmission electron microscopy of therapeutic effects of noble metal nanoparticles in experimental acute inflammation. *RSC Adv.* **2015**, *5*, 67435–67448. [CrossRef]
11. Moldovan, B.; Filip, A.; Clichici, S.; Suharovschi, S.; Bolfa, P.; David, L. Antioxidant Activity of Cornelian Cherry (*Cornus mas* L.) Fruits Extracts and the in Vivo Evaluation of their Anti-inflammatory Effects. *J. Func. Foods* **2016**, *26*, 77–87. [CrossRef]
12. Danila, O.O.; Berghian, A.S.; Dionisie, V.; Gheban, D.; Olteanu, D.; Tabaran, F.; Baldea, I.; Katona, G.; Moldovan, B.; Clichici, S.; et al. The effects of silver nanoparticles on behaviour, apoptosis and nitro-oxidative stress in offspring Wistar rats. *Nanomedicine* **2017**, *12*, 1455–1473. [CrossRef] [PubMed]
13. Ajan, R.; Chandran, K.; Harper, S.L.; Yun, S.-I.; Kalaichelvan, P.T. Plant extract synthesized silver nanoparticles: An ongoing source of novel biocompatible materials. *Ind. Crop. Prod.* **2015**, *70*, 356–373.
14. Prabhu, S.; Poulose, E.K. Silver nanoparticles: mechanism of antimicrobial action, synthesis, medical applications and toxicity effects. *Int. Nano Lett.* **2012**, *2*. [CrossRef]
15. Singh, P.; Kin, Y.J.; Zhang, D.; Yan, D.C. Biological synthesis of nanoparticles from plants and microorganisms. *Trends Biotechnol.* **2016**, *34*, 588–599. [CrossRef] [PubMed]
16. Wei, L.; Lu, J.; Xu, H.; Patel, H.; Chen, Z.S.; Chen, G. Silver nanoparticles: synthesis, properties and therapeutic applications. *Drug Discov. Today* **2015**, *20*, 595–601. [CrossRef] [PubMed]
17. Garcia-Barrasa, J.; Lopez-de-Luzuriaga, J.M.; Monge, M. Silver nanoparticles: synthesis through chemical methods in solution and biological applications. *Cent. Eur. J. Chem.* **2011**, *9*, 7–19. [CrossRef]
18. Sharma, V.K.; Yngard, R.A.; Lin, Y. Silver nanoparticles: green synthesis and their antimicrobial activities. *Adv. Colloid Interface Sci.* **2009**, *145*, 83–96. [CrossRef] [PubMed]
19. Wang, J.; Shan, A.; Liu, T.; Zhang, C.; Zhang, Z. In vitro immunomodulatory effects of an oleanolic acid-enriched extract of *Ligustrum lucidum* fruit (*Ligustrum lucidum* supercritical CO_2 extract) on piglet immunocytes. *Int. Immunopharmacol.* **2012**, *14*, 758–763. [CrossRef] [PubMed]
20. Yin, T.K.; Wu, W.K.; Pak, W.F.; Ko, K.M. Hepatoprotective action of an oleanolic acid-enriched extract of Ligustrum lucidum fruits is mediated through an enhancement on hepatic glutathione regeneration capacity in mice. *Phytother. Res.* **2001**, *15*, 589–591.
21. Lee, S.I.; Oh, S.H.; Park, K.Y.; Park, B.H.; Kim, J.S.; Kim, S.D. Antihyperglycemic effects of fruits of privet (*Ligustrum obtusifolium*) in streptozotocin induced diabetic rats fed a high fat diet. *J. Med. Food* **2009**, *12*, 109–117. [CrossRef] [PubMed]
22. Lin, H.M.; Yen, F.L.; Ng, L.T.; Lin, C.C. Protective effects of *Ligustrum lucidum* fruit extract on acute butylated hydroxytoluene-induced oxidative stress in rats. *J. Ethnopharmacol.* **2007**, *111*, 129–136. [CrossRef] [PubMed]
23. Jeong, J.C.; Kim, J.W.; Kwong, C.H.; Kim, T.H.; Kim, Y.K. *Fructus ligustri lucidi* extracts induce human glioma cells death through regulation of Akt/mTOR pathway in vitro and reduce glioma tumor browth in U87MG xenograft mouse model. *Phytother. Res.* **2011**, *25*, 429–434. [CrossRef] [PubMed]
24. Zhang, J.F.; He, M.L.; Dong, Q.; Xie, W.D.; Chen, Y.C.; Lim, M.C.; Leung, P.C.; Zhang, Y.O.; Kung, H.F. Aqueous extract of fructus Ligustri lucidi enhances the sensitivy of human colorectal carcinoma DLD-1 cells to doxorubicin-induced apoptosis via Tbx3 suppression. *Integr. Cancer Ther.* **2011**, *10*, 85–91. [CrossRef] [PubMed]

25. Hu, B.; Du, Q.; Deng, S.; An, H.-M.; Pan, C.-H.; Shen, K.-P.; Xu, L.; Wei, M.-M.; Wang, S.-S. *Ligustrum lucidum* Ait. fruit extract induces apoptosis and cell senescence in human hepatocellular carcinoma cells through upregulation of p21. *Oncol. Rep.* **2014**, *32*, 1037–1042. [CrossRef] [PubMed]
26. Zeisser-Labouebe, M.; Lange, N.; Gurniy, R.; Delie, F. Hypericin-loaded nanoparticles for the photodynamic treatment of ovarian cancer. *Int. J. Pharm.* **2006**, *326*, 174–181. [CrossRef] [PubMed]
27. Iravani, S.; Korbekandi, H.; Mirmohammadi, S.V.; Zolfaghari, B. Synthesis of silver nanoparticles: Chemical, physical and biological methods. *Res. Pharm. Sci.* **2014**, *9*, 385–406. [PubMed]
28. Singleton, V.L.; Orthofer, R.; Lamuela-Raventós, R.M. Analysis of total phenols and other oxidation substrates and antioxidants by means of Folin-Ciocalteu reagent. *Met. Enzymol.* **1999**, *299*, 152–178.
29. Arnao, M.B.; Cano, A.; Acosta, M. The hydrophilic and lipophilic contribution to total antioxidant activity. *Food Chem.* **2001**, *73*, 239–244. [CrossRef]
30. Kiss, A.K.; Mank, M.; Melzig, M.F. Dual inhibition of metallopeptidases ACE and NEP by extracts and iridioids from *Ligustrum vulgare* L. *J. Ethnopharmacol.* **2008**, *120*, 220–225. [CrossRef] [PubMed]
31. Poopathi, S.H.; De Britto, L.J.; Praba, V.L.; Mani, C.; Praveen, M. Synthesis of silver nanoparticles from *Azadirachta indica*—A most effective method for mosquito control. *Environ. Sci. Pollut. Res.* **2015**, *22*, 2956–2963. [CrossRef] [PubMed]
32. Ashokkumar, S.; Ravi, S.; Kathiravan, V.; Velmurugan, S. Rapid biological synthesis of silver nanoparticles using *Leucas martinicensis* leaf extract for catalytic and antibacterial activity. *Environ. Sci. Pollut. Res. Sci.* **2014**, *21*, 11439–11446. [CrossRef] [PubMed]
33. Sanchez, G.R.; Castilla, C.L.; Gomez, N.B.; Garcia, A.; Marcos, R.; Carmona, E.R. Leaf extract from the endemic plant *Peumus boldus* as an effective bioproduct for the green synthesis of silver nanoparticles. *Mater. Lett.* **2016**, *183*, 255–260. [CrossRef]
34. Elemike, E.E.; Onwudiwe, D.C.; Mkhize, Z. Eco-friendly synthesis of AgNPs using *Verbascum thapsus* extract and its photocatalytic activity. *Mater. Lett.* **2016**, *185*, 452–455. [CrossRef]
35. He, Y.; Li, X.; Wang, J.; Yang, Q.; Yao, B.; Zhao, Y.; Zhao, A.; Sun, W.; Zhang, Q. Synthesis, characterization and evaluation cytotoxic activity of silver nanoparticles synthesized by Chinese herbal *Cornus officinalis* via environment friendly approach. *Environ. Toxicol. Pharmacol.* **2017**, *56*, 56–60. [CrossRef] [PubMed]
36. Dhand, V.; Soumya, L.; Bharadwaj, S.; Chakra, S.; Bhatt, D.; Sreedhar, B. Green synthesis of silver nanoparticles using *Coffea Arabica* seed extract and its antibacterial activity. *Mater. Sci. Eng. C* **2016**, *58*, 36–43. [CrossRef] [PubMed]
37. Sun, Q.; Cai, X.; Li, J.; Zheng, M.; Chen, Z.; Yu, C.-P. Green synthesis of silver nanoparticles using tea leaf extract and evaluation of their stability and antibacterial activity. *Colloids Surf. A Physicochem. Eng. Asp.* **2014**, *444*, 226–231. [CrossRef]
38. El Badawy, A.M.; Luxton, T.P.; Silva, R.G.; Scheckel, K.G.; Suidan, M.T.; Tolaymat, T.M. Impact of environmental conditions (pH, ionic strength and electrolyte type) on the surface charge and aggregation of silver nanoparticles suspensions. *Environ. Sci. Technol.* **2010**, *44*, 1260–1266. [CrossRef] [PubMed]
39. Gengan, R.M.; Anand, K.; Phulukdaree, A.; Chuturgoon, A. A549 cell line activity of biosynthesized silver nanoparticles using Albizia adianthifolia leaf. *Colloids Surf. B* **2013**, *105*, 87–91. [CrossRef] [PubMed]
40. Ahmed, K.B.R.; Nagy, A.; Brown, R.P.; Zhang, Q.; Malhan, S.G.; Goering, P.L. Silver nanoparticles: significance of physico-chemical properties and assay interference on the interpretation of in vitro cytotoxicity studies. *Toxicol. Vitr.* **2017**, *38*, 179–192. [CrossRef] [PubMed]
41. Fahrenholtz, C.D.; Swanner, J.; Ramirez-Perez, M.; Singh, R.N. Heterogeneous response of ovarian cancer cells to silver nanoparticles as a single agent and in combination with cisplatin. *J. Nanomater.* **2017**. [CrossRef] [PubMed]
42. Young, X.; Gondikas, A.P.; Marinakos, S.M.; Auffan, M.; Liu, J.; Hsu-Kim, H.; Meyer, J.N. Mechanism of silver nanoparticle toxicity is dependent on dissolved silver and surface coating in *Caenorhabditis elegans*. *Environ. Sci. Technol.* **2012**, *46*, 1119–1127. [CrossRef] [PubMed]
43. Lakshmanan, G.; Sathiyaseelan, A.; Kalaichelvan, P.T.; Murugesan, K. Plant mediated synthesis of silver nanoparticles using fruit extract of *Cleome viscosa* L.: Assessment of their antibacterial and anticancer activity. *Karbala Int. J. Mod. Sci.* **2018**, *4*, 61–68.

© 2018 by the authors. Licensee MDPI, Basel, Switzerland. This article is an open access article distributed under the terms and conditions of the Creative Commons Attribution (CC BY) license (http://creativecommons.org/licenses/by/4.0/).

MDPI
St. Alban-Anlage 66
4052 Basel
Switzerland
Tel. +41 61 683 77 34
Fax +41 61 302 89 18
www.mdpi.com

Nanomaterials Editorial Office
E-mail: nanomaterials@mdpi.com
www.mdpi.com/journal/nanomaterials

www.ingramcontent.com/pod-product-compliance
Lightning Source LLC
LaVergne TN
LVHW071956080526
838202LV00064B/6763